Pursuing Livelihoods, Imagining Development

Smallholders in Highland Lampung, Indonesia

Asia-Pacific Environment Monograph 9

Pursuing Livelihoods, Imagining Development

Smallholders in Highland Lampung, Indonesia

Ahmad Kusworo

Australian
National
University

PRESS

ANU PRESS

Published by ANU Press
The Australian National University
Canberra ACT 0200, Australia
Email: anupress@anu.edu.au
This title is also available online at: http://press.anu.edu.au/

National Library of Australia Cataloguing-in-Publication entry

Author: Kusworo, Ahmad, author.

Title: Pursuing livelihoods, imagining development : smallholders in Highland Lampung, Indonesia / Ahmad Kusworo.

ISBN: 9781925021479 (paperback) 9781925021486 (ebook)

Subjects: Farms, Small--Indonesia--Lampung.
          Village communities--Economic aspects--Indonesia--Lampung.
          Agriculture and state--Indonesia--Lampung.
          Agriculture--Economic aspects--Indonesia--Lampung.
          Land use, Rural--Indonesia--Lampung.
          Lampung (Indonesia)--Social conditions.
          Lampung (Indonesia)--Social life and customs.

Dewey Number: 338.1609598

Cover design and layout by ANU Press

Cover image: A coffee harvester in Gunung Terang village, photo courtesy of the author

Contents

List of Tables

List of Maps

List of Plates

For Nana, Alia, and Nabila

Foreword

James J. Fox

A strategic perspective is the key to shaping a well-focused ethnography. In this volume, *Pursuing Livelihoods, Imagining Development,* Ahmad Kusworo has developed a perspective that allows his analysis to shift effectively from province to region to village area, at each stage enhancing the overall argument.

At the provincial level, this ethnography presents a superb examination of the transformation of Lampung in Sumatra. Within the context of this historical transformation, the ethnography shifts focus to the specific transformation of a frontier region of Lampung and then concentrates on a fine-grained examination of local rural development – how cultivators of mixed backgrounds flexibly organized in households and attuned to new opportunities manage to pursue a range of livelihood strategies. The result is a sophisticated, nuanced ethnography that provides an exceptional portrait of smallholder production.

Lampung is an area of Indonesia that has not been given the attention that it deserves. In the colonial period, Dutch administrators viewed Lampung as an empty land that they could fill with people from Java. The province became the earliest historical target for migration and continued as such through much of the New Order. Over time, there was far more spontaneous migration than assisted transmigration, and the local indigenous population — though swamped by those from outside — took on many of the same economic pursuits as the incoming groups. The result is a complex creation of diverse settlements pursuing multiple livelihoods. The author aptly describes this population as "a multi-ethnic middle peasantry" whose efforts have resulted in successes as well as failures. This volume traces the effects of the erratic national policies and fluctuating commodity prices that have affected opportunities and the way of life in the region. It provides a convincing picture of 'development' as it is imagined and experienced in the local context.

To do this kind of nuanced research requires time and a solid background in Lampung's diverse cultural traditions. Ahmad Kusworo received his first degree in agriculture from the University of Lampung and was engaged in research in the province with Friends of Nature and Environment (WATALA), and ICRAF/ World Agroforestry Centre before beginning his degree program at ANU. Almost a decade before he began his fieldwork, he visited the area of his research in the company of friends and activists from various NGOs. Since completing his PhD, Ahmad Kusworo has worked for WFF Indonesia and UNDP Indonesia as well as being the Indonesian Research Coordinator for ANU Crawford School's Australia Indonesia Governance Research Partnership. He is currently a Technical Advisor for the Indonesia Programme at Fauna & Flora International.

This is an engaging ethnography that is particularly appropriate for the Asia Pacific Environment Monograph (APEM) series. It is an ethnographic document of a particular place and a particular time but the issues it tackles are perennial and persistent.

Acknowledgements

This book is the polished version of my PhD thesis submitted in 2004 at the Department of Anthropology, Research School of Pacific and Asian Studies (RSPAS) at The Australian National University. I'd like to thank members of the department: James J. Fox, Andrew Walker, Lesley Potter, and Andrew McWilliam who have supervised, advised, and helped me throughout the process of preparation, fieldwork research, and writing of the thesis. I thank colleagues from the World Agroforestry Centre (ICRAF) and Friends of Nature and the Environment (WATALA) for hosting me during the fieldwork research. The Ford Foundation and RSPAS generously provided funds for my PhD program.

Michael Dove (Yale), Nancy Peluso (UC Berkeley), and Carol Warren (Murdoch University) have kindly examined the thesis and offered valuable comments and suggestions for the publication of this monograph.

I thank James J. Fox for connecting me with people at the Resource Management in Asia-Pacific (RMAP) Program at ANU. Colin Filer, Mary Walta, and Dana Rawls provided support with editing and publishing this monograph.

Glossary

Acronyms

ABRI (Angkatan Bersenjata Republik Indonesia)	Indonesian Armed Forces
babinsa (bintara pembina desa)	military personnel posted in village
BRN (Biro Rekonstruksi Nasional)	National Reconstruction Bureau
Corp Tjadangan Nasional (CTN)	National Reserve Corp
DI/TII (Darul Islam/Tentara Islam Indonesia)	the House of Islam/Indonesian Islamic Army
DPRD (Dewan Perwakilan Rakyat Daerah)	Regional House of Representatives
Golkar (Golongan Karya)	Functional Party
HKm (Hutan Kemasyarakatan)	a version of community/social forestry
HTI (hutan tanaman industri)	industrial forestry plantation
ICRAF	International Centre for Research in Agroforestry/ World Agroforestry Centre
IDT (Instruksi Presiden Desa Tertinggal)	Presidential Instructions on Neglected, Left-Behind Villages
Koramil (Komando Rayon Militer)	sub-district military post
LPMP (Lembaga Pemberdayaan Masyarakat Pekon)	village council for community empowerment
NGO	Non-government organisation
ORSTOM	Institut Français de Recherche Scientifique pour le Développment en Coopération
PBB (Partai Bulan Bintang)	Moon Star Party
PDIP (Partai Demokrasi Indonesia Perjuangan)	Indonesian Democratic Party of Struggle
PKB (Partai Kebangkitan Bangsa)	Nation Awakening Party
PKI (Partai Komunis Indonesia)	Indonesian Communist Party
PPP (Partai Persatuan dan Pembangunan)	Unity and Development Party

TGHK (Tata Guna Hutan Kesepakatan)	Agreed Forest Land Use Plan
WATALA (Keluarga Pencinta Alam dan Lingkungan Hidup)	Friends of Nature and Environment
WALHI (Wahana Lingkungan Hidup Indonesia)	Indonesian Forum for Environment
YLBHI (Lembaga Bantuan Hukum Indonesia)	Indonesian Legal Aid Foundation

Terms

Indonesian/local	English
adat	customary practices
alang-alang	*Imperata* grassland
agen bis	bus ticketing agent
arisan	rotational credit and saving group
babinsa (bintara pembina desa)	military personnel posted in village
bandar	(local) political representative
belukar	bush, fallow field
bawon, ngepak/ceblok	share harvest arrangement in ricefield cultivation
boschwezen (BW) [Dutch]	forestry zones
bujang, nggarap	contract laborer paid with share of harvest in coffee field
buay	patrilineal grouping
buruh tani	agricultural labourers
bupati	district head
cakak padun	ritual feast for noble title inauguration
camat	sub-district head
cangkai	sub-lineage
cicilan	down payment, installment
controleur [Dutch]	government officer
cultuurgebied [Dutch]	plantation belt
dadap, chinkareen	*Erythrina spp*, coffee/pepper shade trees
dangdut	Malay reggae music

demang	non-indigenous worker
desa	administrative village
dukungan masyarakat	mass/popular support
dusun	sub-village
gabah	unhusked rice
gerombolan	band of men
gotong royong, kerja bhakti	mutual help, mutual aid, communal work, non-paid labour
haji	pilgrimage to Mecca
hutan kemasyarakatan/HKm	community forestry
hutan tanaman industri/HTI	industrial forestry plantation
janggolan	tax (in cash or kind)
jenjem	political representative
jurai	lineage
kabupaten	district
kakak adik	siblings
kaleng	tin can/container
kawasan hutan lindung	protection forest zones
kawasan hutan negara	state forestry zones
kebun	smallholder gardens
kecamatan	administrative sub-district
kemajuan	progress
kepala desa, peratin	village head
kepala tebang	chief of clearing
kepala negeri	chief of merged *marga*
kepala dusun/suku, pemangku	head of hamlet/sub-village
kewedanan	sub-district administration
komisaris	commissioner
kopassus (komando pasukan khusus)	army special squad
kota administratif	municipality
koramil (komando rayon militer)	local military post
krismon (krisis moneter)	monetary crisis, 1997–98
kurang berkembang	under-developed
ladang	swidden fields, upland rice fields
madrasah	Islamic school
marga	clan/shire/large socio-political grouping

maro, garap	sharecropping
minggat	leaving without saying
musiman	usury, debt with very high interest
mushalla	small mosque
numpang	using somebody else's property for free; temporary village resident
nangkit	inheritance of property by eldest son when families have no daughters
negeri	administrative merging of several *marga* into a sub-district
numpang	using a plot of land for free; to borrow
nuwo	house
nyusuk	group of families working together to find new land to farm
ojek	motorbike taxi
onderafdeeling [Dutch]	territorial administrative sub-division
orang	person, people
padi ladang, padi darat	upland rice
pam (petugas keamanan) swakarsa	civilian forest rangers
pak	father or sir
paksi	(local) political representative/leader
pembangunan	development
pekon	village
pencak silat	martial arts
pengajian	Qur'an reading
penyimbang	village level leaders
pepadun	seat used in title-granting feast ceremony
perambah hutan	forest squatters, encroachers
pesantren	Islamic boarding school
pesirah	government appointed *marga* head
proatin	village assembly
pusaka	heirloom
puyang	great-grandparent, ancestor
ratu	lord (king, queen)
rebana	tambourine

reformasi	reformation
selametan, syukuran	ritual feast
sawah	irrigated rice field
sedekah	ritual feast
sedekah pusaka	ritual feast to commemorate ancestors
sekolah menengah pertama	junior high school
singkuh sinduh	polite/approriate behaviour between marriagable men and women
suku	patrilineal groupings, hamlet
sumbangan	gift, donation
talang	agricultural field
tambak	fish or shrimp pond
tanah kawasan (hutan negara)	government designated forestry zone, state forest land
tanah marga	non-state land
tegalan	dry field plot
transmigrasi lokal (translok)	local transmigration program
tunggu tubang	inheritence of property by eldest daughter
umbul	flag or banners
ulasan	token payment
upahan, harian	daily wage labour
wedana	district administrator
warung	stall, kiosk, shop
yasinan	Qu'ran recitings
ziarah	visit to a sacred place

1. Introduction

This monograph examines the ways in which people experience 'development' and how it can shape and influence their lives. The book will explore forces that can drive change and some consequences of these forces, including ways people cope with change. The region of focus is Lampung, the southernmost province of Sumatra, Indonesia. The approach explores local understandings within a local and regional context.

My exploration begins at the provincial level, moves to one of the province's highland regions, and concludes at a selected highland village. The increasing narrowness of geographical focus provides an opportunity to look at development on a variety of scales. The selected approach is an attempt to overcome what Eric Wolf (1982: 13) has called 'the false confidence' of micro-level ethnography. Similarly, the approach is employed to avoid treating 'societies … [or] villages … as if they were the islands unto themselves, with little sense of the larger systems of relations in which these units are embedded' (Ortner 1984: 142).

Imagining Development and Change

Two widely held views of post-colonial development are that (1) it has failed to deliver its stated objectives, and/or (2) it has been rejected by its intended beneficiaries. From these viewpoints, development can be seen as superceding colonialism as a new mode of domination and exploitation (see Sachs 1992; Ferguson 1994; Escobar 1995).

Escobar (1995) has identified development as a regime of knowledge embedded in global asymmetrical power relations. In his view, development concerns a set of ideas and practices to bring or deliver 'progress'. These ideas and practices are produced by and serve the interests of the First World (the North) and are applied to the Third World (the South). Consequences have been continued domination by the First World of the Third World, accompanied by processes of underdevelopment and resistance to development in the South. Continued poverty and environmental degradation have been the legacy of this structure of relations.

In a similar vein Ferguson (1994), basing his ethnographic study in Lesotho, identified what he considered to be 'real' effects of development. According to his account, the applied development failed to improve people's livelihoods, primarily as a result of offering technical solutions to non-technical problems.

However, the real effect of development was the expansion of state power where development projects became the primary tool to improve the welfare of the people. Another real effect of development projects — that are planned, funded, and implemented by numerous international development agencies — has been the emergence of a global development industry (ibid.).

Hobart (1993) attributed the failure of development to the growth of 'ignorance', positioning development as a key element in determining global post-colonial relations. He saw the production and reproduction of development packages as often guided by principles of Western scientific knowledge. For Hobart, this Western scientific logic and rationality was incompatible with and in opposition to local indigenous knowledge. He suggests that it is little wonder that development packages have often ended in failure and describes development practitioners as being ignorant of local knowledge and continuing to apply inappropriate models based on Western scientific knowledge. The growth of such ignorance is believed to have kept development businesses running. Hence, Hobart concludes that processes of 'under-development' have continued.

Grillo (1997) offers a different perspective on what he understands as the 'real' effects of development. He regards Hobart's viewpoint and that of others (Ferguson 1994; Escobar 1995) as largely representing a 'myth'. Grillo suggests that development is (poorly) represented as a 'monolithic enterprise, heavily controlled from the top, convinced of the superiority of its own wisdom and impervious to local knowledge' (Grillo 1997: 20). He argues that Hobart and others only permit developers, victims, and/or resistors to be involved in development and ignores other responses, agendas, and actors. Moreover, the myth oversimplifies the situation and positions the dominant power as an easy target. It fails to capture the multiple, diverse voices and realities embedded in the processes of planned change and transformation. Far from complete, static and impermeable structures, Grillo suggests that both Western scientific and indigenous knowledge continue to change and to be exchanged. As such, actors in development must adjust their perspectives and positions as circumstances change.

There are at least two key approaches in studying development, neither of which is mutually exclusive. The first is through observing and interpreting the ways people are affected by and/or react to development practices. The second is by studying development in the context of the expansion of power. With respect to these two approaches, James Scott's work is of particular importance. Scott (1998) has approached development using both of these methods. His arguments are grounded in powerful concepts such as 'weapons of the weak' (Scott 1985) and 'simplification and legibility' (Scott 1998).

Scott (1985) used ethnographic materials to demonstrate ways in which peasants in a village in Malaysia experienced and reacted to the Green Revolution, the increase in agricultural production due to improved technology that occurred from the early 1940s to the late 1970s. He argues that Green Revolution initiatives in relation to rice cultivation (for example, improved varieties, double cropping and engine-powered harvesters) made the rich richer while the poor remained poor. The poor used 'everyday forms of resistance' — including a war of words, boycotts, disguised strikes, and petty theft — as 'weapons' in their class struggle against the rich and indirectly, against the state (ibid.). Scott claimed the poor peasants' greatest accomplishment was to delay the complete transformation to capitalist forms of production, which was the aim of the agricultural development policy implemented by the Malaysian state.

In a later work, Scott's concept of 'simplification and legibility' (1998) provided an explanation of why development schemes for improvement of human conditions have failed. According to Scott, 'the legibility of a society provides the capacity for large scale social engineering' and examples of development initiatives from around the globe — from 'scientific' forestry, agricultural development and city planning, to Soviet and African socialism — were painstakingly analysed. The failure of these schemes was attributed to expanding state power in order to control resources and people which was achieved by 'simplifying' complex, local, social practices from the 'centre' and above, enabling those in power to record, monitor, and manipulate their subjects. In the process, local knowledge and know-how were ignored within the simplified administrative grid of formal state observations.

In criticising this analysis, Ortner focused on the problem of locating resistance in its everyday forms. She raised the question of what is or is not resistance. 'When a poor man steals from a rich man, is this resistance or simply a survival strategy?' (Ortner 1995: 175). She argues that:

> resistance … highlights the presence and play of power in most forms of relationship and activity … [but] we are not required to decide once and for all whether any given act fits into a fixed box called resistance …. [T] he intentionalities of actors evolve through praxis, and the meanings of acts change, both for the actors and for the analysts (ibid.).

And the elements that need to be emphasised include:

> the ambiguity of resistance and the subjective ambivalence of the acts for those who engage in them … [because] in a relationship of power, the dominant often has something to offer and sometimes a great deal (though always of course at the price of continuing power). The subordinate thus has many grounds for ambivalence about resisting the relationship.

> Moreover, there is never a single, unitary, subordinate … in the simple
> sense that subaltern groups are internally divided … [into various]
> forms of difference and that occupants of differing subject positions will
> have different, even, opposed, but still legitimate, perspectives on the
> situations (ibid.).

However, within the complexity of resistance and non-resistance (cooperation,
reciprocity, and harmony) there is the tendency to overlook the latter. In this
regard, Pelzer White advises that 'we must add an inventory of "everyday
forms of peasant collaboration" to balance our list of "everyday forms of
peasant resistance" — both exist, both are important' (White 1986, quoted in
Ortner 1995: 176). In a similar vein, Brown somewhat exaggeratedly points out:

> [H]uman institutions … [such as] family, organisations, and systems of
> production doubtless impose forms of subjugation, [but] they are also
> institutions that enable. Without them society would cease to exist, and
> with it, the capacity of human beings to survive (Brown 1996: 734).

Like the concept of 'legibility', Scott's 'everyday forms of resistance' place
the state and the people in oppositional frameworks within this development
context. In situations where development brings mixed results rather than only
failure and resistance, alternative conceptual tools are needed. In dealing with
the initiatives of development and its concomitant changes, people's responses
or strategies involve competition, accommodation, and compliance as well as
resistance.

In the modern Indonesian uplands, as Li (1999a: xvii) explains, the state's
primary concern 'has been to bring order, control and "development" to upland
regions while deploying upland resources to serve national goals'. Key state
initiatives in the Indonesian uplands are territorialisation and development
(Li 1999b). Through territorialisation 'modern states divide their territories
into complex and overlapping political and economic zones, rearrange people
and resources within this units and create regulations delineating how and by
whom these area can be used' (Vandergeest and Peluso 1995: 387).

State power in the Indonesian uplands has been directed towards achieving
greater control over resources and people. A large portion of the land is classified
as state forest for lease to logging and forest plantation companies. The net effect
is to prohibit access to local people and transform them into labourers. Logged-
over lands are then 'developed' into large-scale plantations by state and private
companies or alternatively, designated as transmigration sites that 'promote
economic growth while also bringing political and administrative order to
peripheral areas' (Li 1999b: 15–16). A less aggressive initiative to intensify state
control over people and resources has been accomplished by regularising the

spontaneous incursion of migrants into frontier zones (ibid.: 17). Once newcomers have been organised into administrative units (*desa*), their daily activities can be monitored and regulated through the various village committees and institutions specified by law. This initiative is made easier as newcomers want and need to be enmeshed in state systems in order to claim their place as citizens and as clients of state officials and institutions. Their eagerness to be welcomed into the fold could potentially legitimise their presence and consolidate their hold over resources.

At the heart of development relations lies a tension between 'centres' and 'peripheries'. In this context, Tania Li introduced the concept of 'relational formations' of social marginality (Li 1999b, 2001). Marginality emerges from ongoing centre and periphery relations, rather than from resistance of the periphery toward the centre or the absence of centre and periphery relations. Indonesian upland communities often depicted as geographically isolated and socially marginal, such as the Meratus Dayak in Borneo (Tsing 1993) and the Lauje people in Sulawesi (Li 2001), were arguably created through the engagement of local tribes with pre-colonial courts, colonial administrations and post-colonial regimes. In pre-colonial and colonial periods, relations took the form of rule and trade; in post-colonial times 'development' is the leitmotif. Li explains:

> [L]ike the Lauje, the Meratus practice shifting cultivation and continue to live and move about in ways that are illegible to the government administrators nominally responsible for them. Yet they are not an autonomous group resisting outside authority (ibid.: 44).

Their marginality was developed in dialogue with state formations. '[T]heir lifeways are formed not outside state agendas but relationally, in and through them' (ibid.).

Li (2001) also contends that in cases like the Lauje of Sulawesi, rule and trade relations enabled the centre to control the people and exploit local resources (forest products, agricultural commodities, and labour) in the absence of legibility (maps, statistics, and monitoring). In the context of the failure of Indonesia's New Order rural development programs, Dove and Kammen (2001: 633) suggest that the state produced illegibility as much as legibility. Illegibility is not an accidental product of weak governance, but may form a strategy by political central elites for political and economic purposes. State-based appropriation and exploitation of economic resources are facilitated in the absence of clearly defined local rights and practices.

Dove and Kammen (2001), also working within the framework of relations between the centre and periphery, examined development in terms of resource flows in everyday practices of development in Indonesia's New Order era (1966–98). They suggested that:

> [T]here were two co-occurring models of development: an official one and a `vernacular' one. The former represents a formal, uniform, and idealised vision of what the state professed, whereas the latter represents a more nuanced, normative, and conflicting vision of what state agents actually strove for. The vernacular model is an intentional one: it was the product not of accident but institutionalised values and desires. (ibid.: 633)

As opposed to official models where development is supposed to promote the flow of resources from centre to periphery, vernacular models of development enabled the centre to block the flow of resources from the centre to periphery and in fact reverse the flow by extracting resources from the periphery. Contract farming on rubber cultivation, for example, was heralded as a way to provide assistance to the smallholders. In reality, the estate companies used contract farming on sugar cane as a means for sugar companies to extract resources from local smallholders while preventing the flow of other resources back to them. The types of resources that 'were allowed to proceed unhindered down and out from the centre were those that central elites did not want' (ibid.: 626–7). Examples of resources that were successfully transferred to marginalised people were Social Department programs and family planning services whose resources were modest. However, these agencies promulgated a view that the receiving areas were deficient and undeveloped, thereby justifying an increase in state intervention. Another resource that central elites allowed to flow were allotments in the transmigration program. 'According to the official model of development, the state gave valuable resources to marginal groups; according to [the] vernacular model, it gave value-less resources to them' (ibid.: 627).

Locating the state/centre and people/periphery perspective relationally, the concepts of relational formation and the vernacular model of development can be applied strategically to analyse the formation of social marginality (Li 1999a, 2001) and the failure of development schemes (Dove and Kammen 2001). In this book, the concepts of relational formation and the vernacular model of development are combined and modified to analyse situations where state/centre and people/periphery relations do not necessarily lead to marginality and development failures. This study explores how people in geographically marginalised areas position themselves within the orbit of state power in order to promote resource flows from the centre to the periphery, while restricting resource extraction from the periphery to the centre.

To look only at the ill-effects of development risks overlooking its manifold benefits. In relation to the impacts and effects of development in Southeast Asia, Rigg contends that:

> [I]t is hard to think of one indicator of human well-being that has not improved during the course of modernisation over the last half century. It is notable that those countries which have experienced sustained stagnation or decline in such indicators are those that have experienced an absence of development as modernisation …. [D]evelopment has led to real, substantial and, in some cases, sustained improvements in human well being …. Nor can this be rejected as a case of the benefits accruing to just a small segment of the population, leaving the majority mired in poverty …. [I]mprovements in livelihood have been broadly based, even if they have not been equally distributed (Rigg 2003: 328–9).

Development is 'as much a fact of everyday life for most people of the world as other kinds of overarching frameworks of assumption and action' (Croll and Parkin 1992, cited in Grillo 1997: 1). Pigg (1992), discussing examples from Nepal, goes on to assert that development connects villagers, the urban elite, national political institutions, international development agencies, and representatives of the Third World in the West. 'Everyone wants a piece of the development pie' (Pigg 1992: 511).

Smallholders, Production and Differentiation

Fundamental to an understanding of development and change are the ways rural populations reproduce their modes of livelihood. The people discussed in this study are predominantly smallholder farmers. The argument advanced is that flexibility in the social organisation of agricultural production and in the use of available resources to respond to constraints and opportunities is the key to the persistence of smallholder farming (as a system of agricultural production) and the smallholding tradition (as a social–agrarian structure).

People discussed in the present study accord with Netting's definition of smallholders as:

> rural cultivators practicing intensive, permanent, diversified agriculture on relatively small farms in areas of dense population. The family household is the major corporate social unit for mobilising agricultural labor, managing productive resources, and organising consumption. Smallholders have ownership or other well-defined tenure rights in land that are long term and often heritable (Netting 1993: 2).

Netting's classification thus excludes:

> shifting cultivators practicing long fallow or slash and burn farming where land is still plentiful and population density low, as in some parts of the humid tropic today; … herders whether they are nomadic pastoralists of east Africa or the ranchers of Texas; … and the farming systems of dry monocropping of wheat, sugar estates, cotton plantation with slaves, or California agribusiness (ibid: 2–3).

The argument that Netting advances for the persistence of smallholder household farming is that 'intensive agriculture by landowning smallholder households is economically efficient, environmentally sustainable and socially integrative' (ibid: 27).

One of the key characteristics of smallholder production is the superiority of household labour compared to communal labour in collective farming, or to hired labour in capitalist farming (ibid.). It is the smallholder household members who perform the diverse, skilled, and — in this setting — unsupervised tasks of intensive cultivation.

Population pressures and the market are often implicated as the driving forces of agricultural transformation toward intensive farming. Population growth increases land scarcity and promotes agricultural intensification (Boserup 1965), while markets create demand for cultivated commodities (Netting 1993, Brookfield 2000). Land abundance, along with the market attraction of rubber and labour shortages, has caused the indigenous people in Kalimantan to cultivate extensive rice swiddens and rubber gardens (along with other tree crops). With labour abundance and shrinking landholdings, rural populations in Java practise intensive irrigated rice cultivation (Dove 1986). With respect to economic efficiency of small-scale agricultural production, Dove notes that the production of intensive irrigated rice cultivators is significantly different from swidden cultivators. Where land availability is a constraint, intensive irrigated rice cultivation is aimed at a high return, namely production per unit of land. In swidden cultivation, which is characteristically constrained by labour shortages, farming is oriented towards a high return to labour, or production per unit of labour.

Brookfield (2000) emphasises capital and skills as the key elements of agricultural transformation besides labour. However, increases in productivity may not necessarily follow an increase in labour input. Conversely, there are cases where increases in production can actually reduce the demand for labour. Here investment in working capital, such as tools and animals, may be more closely linked to increases in production. Farmers' skills are usually thought of in terms of agro-technical skills, but organisational skills are also often very important.

Although agricultural transformations can be triggered by various factors such as new technology, expanding commercialisation, and/or state interventions, the real foundation of such transformations is the skill of small farmers to organise their land, workforce and resources. Brookfield goes on to argue that the key factor in maintaining small farms' ecological and production sustainability is agro-diversity, meaning a diversity of plant and animal species being cultivated. This approach requires special farming methods and labour organisation together with a deep knowledge of ecology and technology. Having the capability to respond and adapt to market opportunities is also crucial.

Flexibility in '[t]he ability to use different resources, and employ different strategies for making a living' (Brookfield 2001: 187) is another key aspect for understanding agricultural transformations. Agricultural transformations can occur through intensification, meaning that improved productivity can be achieved through increased labour input, or 'dis-intensification' where increased productivity can be achieved through improved farming skills and techniques leading to a reduction of labour. In many cases, Brookfield argues, increasing production 'involve[s] new skills in [the] use of "dynamic" land, and both agricultural and non-agricultural opportunities, and not increased inputs into any constant land or … increased current inputs of any kind except of management skills' (ibid.: 189).

More often than not, smallholder farmers' commodity market productions are made possible due to the incorporation of non-market capital. In the Sulawesi highlands, for example, one strategy was to use non-market inputs such as mutual labour assistance to pursue market relations without which the production of rice for the market is difficult or may not even be possible (Schrauwers 1999). Similarly, for Minangkabau smallholder farmers (Khan 1999), the main inputs for production such as labour, land, and capital were obtained largely through non-market capital. Access to land, for example, was obtained through inheritance, sharecropping, and squatting on forest reserves and plantations. In the production of rubber in Riau, smallholders retain their traditional elements of the farming system such as cultivation of jungle rubber, customary (*adat*) and communal land ownership and, wherever possible, subsistence rice farming (Potter and Badcock 2004).

State policies regulating access to upland lands in Indonesia often influence smallholder intensive agricultural practices. A large portion of the Indonesian uplands has been either classified as forest reserves or otherwise granted to plantation companies. Many indigenous peoples in Sumatra, Kalimantan, and Sulawesi have changed their agricultural practices from dry land swidden of rice to tree crop cultivation and managed agro-forests, sometimes with accompanying wet rice (Potter 2001), in response to the loss of land to forest reserves and estate plantations. In the Lauje hills and Lindu areas of the Sulawesi Highlands, the government's inability to control these 'state lands' has enabled the indigenous population and Bugis migrants to use these forests and former swiddens, turning them into intensive cocoa groves (Li 2002).

One analysis of why smallholder traditions have seemingly persisted is conducted by examining agricultural transformations in the form of changes in farming practices. A second line of inquiry examines changes in the social organisation of farming. A key element in the social structure of rural society is rural differentiation.

Ben White defines agrarian or rural differentiation as a process that:

> involves a cumulative and permanent (i.e., non-cyclical, which is not to say that it is never reversible) process of change in the ways in which different groups in rural society — and some outside it — gain access to the products of their own or others' labor, based on their differential control over production resources and often, but not always, on increasing inequalities in access to land (White 1989: 20).

He makes a further distinction:

> between the process of differentiation itself and various aspects of that process which we might call the causes, the mechanisms, and the symptoms or indicators of differentiation. Similarly, any analysis of rural differentiation processes in a specific place and time will have to encompass their contexts (regional, national, political, cultural, etc.) and also the constraints to differentiation (which may originate externally or internally and may affect the pace and form of differentiation) (ibid.: 25–6).

Netting (1993) suggests that smallholder agriculture is akin to gambling, where some players, due to their individual ability, play the game better than others. Differentiation and inequality are inevitable in this circumstance. The state often plays an important role in promoting or constraining differentiation. In the Tengger Highlands in East Java, for example, styles of land distribution by the colonial government led to the emergence of a 'smallholding tradition' (Hefner 1990). In lowland Java, New Order initiatives such as the Green Revolution and the absence of land reform favoured village elites and promoted differentiation (Hart et al. 1989). Li (2002) and Potter and Badcock (2004) suggest that studying agrarian structure is an exploration of human agency. The agrarian structure is the result and medium through which rural people discover and optimise constraints and opportunities in order to obtain the 'good life'.

In lowland rice areas in Java, White and Wiradi (1989) reported that ownership of rice fields was very inequitable and differentiation ensued. On one hand, wealthy households have many avenues for profitable investment and many demands that require non-productive expenditures that compete with the need for land acquisition. On the other hand, the many smaller owners whose agricultural incomes do not provide for production even at minimal levels are able to achieve subsistence incomes without the distress sale of their 'sub-livelihood' plots by participating in a variety of low-return, non-farm activities both inside and outside of the village.

In the Tengger Mountains (Hefner 1990), Malang (Suryanata 1999) and the Sulawesi uplands (Li 2002), wealthy migrants have taken over a large portion of upland food and cash crop fields through buying, renting and/or mortgaging property. In the process they have converted a large number of the local inhabitants into landless labourers. This differs from the situation in Langkat, North Sumatra (Ruiter 1999), where Batak villagers retained their control over smallholder rubber gardens, leaving the Javanese migrant labourers to occupy the lowest of the village's socio-economic strata.

Hefner (1990) has pointed to a distinct rural social group he has interchangeably called the 'middle peasantry' or 'smallholder peasantry' whose ethos and aspirations are to maintain the 'smallholding tradition'. Their persistence is attributed to a desire for social autonomy and their capacity to own land. 'Situated between the more visible agrarian elite and the mass of the poor', Hefner asserts, 'the middle peasantry … received scant comments in many agrarian accounts of agrarian change. Influenced by … [the] vision of social polarisation … scholars assume that middle peasants are doomed to historical oblivion' (ibid.: 154). Villagers in the Tengger Mountains, like rural people elsewhere in Java, were affected by the shrinking landholding pool as it was increasingly taken up as a consequence of emerging national markets and ensuing politics. The villagers, Hefner claimed, acknowledged that there are 'haves' and 'have-nots' but '[they] deny [the] suggestion that there might be an unbridgeable gap between rich and poor' (ibid.). Hefner went on to explain that the middle peasantry in the Tengger Mountain regions:

> is characterised by neither the servile dependence of a dominated underclass nor the collective solidarity romantically attributed to proletarians…. Its social orientation emphasized neither selfless collectivism nor self-possessed individualism. Its animating ethos is an almost-paradoxical mix of self-reliance and communalist commitment. Ideally, in this view, each household guarantees its own welfare (ibid.).

Hefner continues:

> The aspiration of these uplanders … is … [that] one seeks to stand on one's own and not to be ordered about. Only in doing so can one be fully

acknowledged as a member of the community. The simple achievement of respectful standing in a community of brethren is a valued end in its own right (ibid.: 157).

Smallholders in the Lampung highlands are also characterised by the ethos and aspiration stressing social autonomy; that 'one seeks to stand on one's own' and 'each household guarantees it own welfare'. Their aspirations include: having enough money to meet family needs; providing an education for their children; acquiring modern household equipment; improving their housing; and having access to credit. These goals are to be achieved through personal development, 'the development of a person by themselves' (Green 2000: 68). This 'self-development' is pursued within the context of state-led development. For migrant smallholders in the Lampung highlands, state-led development offers resources that have enabled them to achieve self-development goals. They have transformed a forest frontier into a flourishing highland. In the process, as this book argues, they have produced and reproduced the smallholder tradition. It is further argued that their village's social life is organised principally to attract state resources and to reap the benefits of development.

The Field Research

Research for this monograph was conducted between March 2002 and February 2003 when I lived in Sumber Jaya and Way Tenong, two adjoining sub-districts (*kecamatan*) in West Lampung District (*kabupaten*). Most of the time I lived in the village of Gunung Terang in two different homes — first in a Semendo neighbourhood in the main village hamlet of Gunung Terang, and then in the hamlet of Rigis Atas on the slopes of the Bukit Rigis Mountain.

I visited and sometimes stayed for several days in other villages of the region. Friends from WATALA[1] and ICRAF[2] often visited me in the village or invited me to visit their work sites. I also regularly participated in their community meetings and workshops. ICRAF and WATALA have been working in West Lampung District for several years to support negotiation processes between local communities and government agencies on the issues of natural resource management. ICRAF scientists collaborate with various national research institutions and also conduct their own socio-economic, biophysical and policy research in the region. During my stay in Gunung Terang, friends from WATALA and ICRAF conducted community mapping of the village and assisted

---

1   Keluarga Pencinta Alam dan Lingkungan Hidup [Friends of Nature and the Environment], an environmental NGO founded in 1978 by students in the Faculty of Agriculture at Lampung University.
2   The International Centre for Research in Agroforestry, now called the World Agroforestry Centre.

the community group in Rigis Atas with obtaining a community forestry permission contract. Assistance in obtaining such contracts was also given to community groups in other villages across the region.

The 2002–3 fieldwork was not my first visit to the Sumber Jaya and Way Tenong region. Between 1994 and 1995, I visited the region with other friends and NGO activists from WATALA, WALHI[3] and YLBHI[4] at the invitation of the World Bank and PT PLN (Perusahaan Listrik Negara, the state-owned electricity company) to assess the social impact of the construction of the Way Besai Dam and to discuss plans to mitigate these impacts.[5] We were expecting villagers' resistance to this mega project, but to our surprise villagers were receptive and local leaders denied the suggestion that villagers rejected the project. When we pointed out possible hardships for landless and near landless villagers in finding alternative sources of livelihood — as suggested in the environmental impact assessment report — a common response from village leaders was that the project would provide more benefit than harm. A Semendo village leader even stated that to refuse the project was a sin and against their ancestors' wishes. It was said that their ancestors knew and had told them that a big dam would be built in the area.

During the 1994–95 period, the military began operations to destroy smallholder gardens and houses inside the boundary of the state forest zones for replacement by plantation forests. The market villages of Sumber Jaya and Fajar Bulan were transformed into small market towns. The villages were electrified. Along the road, sturdy wooden and brick houses were constructed or refurbished thanks to the rise in prices and production of coffee. Between 1996 and 1998, on my trips to and from Krui I frequently stopped in Fajar Bulan and Sumber Jaya either for a short rest or to meet acquaintances.

Between 1998 and 2000, I conducted a series of fieldwork visits to Sumber Jaya and Way Tenong. I was working for ICRAF and visited different parts of the region with friends from WATALA. We did a general survey of community–forest interactions and household economy. This was the period of *reformasi* (the overthrow of Suharto in 1998 and the demise of the New Order regime), the *El Niño* drought, and the *krismon* monetary crisis in 1997 which led to a dramatic fall in value of the Indonesian currency, all of which were embraced as 'good things' by the people in the region. *Reformasi* was interpreted as granting 'freedom' to reclaim land in forest zones, *El Niño* effectively brought higher

---

3 Wahana Lingkungan Hidup Indonesia [Indonesian Forum for Environment]) is a national NGO with a secretariat in Jakarta and regional secretariats in many Indonesian provinces.

4 Lembaga Bantuan Hukum Indonesia [Indonesian Legal Aid Foundation] has their headquarters in Jakarta and regional offices and posts throughout Indonesia.

5 The project paid relatively high compensation to hundreds of families in Way Petai, Sukapura, and Dwikora whose rice fields and coffee gardens were used for the project. PT PLN also provided credit for the village community groups.

coffee production and made the dried shrubland easy to burn, while the impact of the *krismon* caused a hike in the export price of coffee and brought in a flush of money. The region was flooded with luxury items from wool blankets and electronics to motorbikes and cars.

When I returned to the region in early 2002, the 'good times' of *reformasi* and *krismon* were over. Cars and motorbikes had been sold and many houses that had been under construction were left unfinished. More recent migrants had left the region to return home or had moved on to new frontier zones in the neighbouring province of Bengkulu and elsewhere. The talk among ordinary smallholders in the region changed from aspiring to higher education for the children and sturdy modern houses for themselves to how to provide enough food for their families and sufficient inputs to their diversified agricultural production.

Book Outline

Chapter Two traces the history of Lampung in the twentieth century. The focus of discussion is on the rural areas of the province. Depicted as an 'empty land' in the early 1900s, by the end of the century Lampung was perceived as a province peopled by land-hungry migrants. Colonial and post-colonial initiatives were identified as the driving forces of Lampung's transformation in the twentieth century. Colonial and post-colonial government initiatives aimed at bringing 'progress' to Lampung brought mixed results including rapid growth in agricultural production, the formation of 'wealthy zones' in some areas, and the creation of pockets of poverty in other areas. The chapter explores the ways people in different rural regions of the province have experienced this transformation.

Chapters Three, Four, and Five explore how migrants transformed one of Lampung's 'last frontiers' into one of its highland 'wealthy zones'. The chapters also explore how these migrants shaped their own modes of life in the process. Chapter Three gives an account of the history of the influx of different groups of migrants to settle in the Sumber Jaya and Way Tenong region. Although the bulk of these migrants migrated 'spontaneously', they were heavily integrated into the planned development framework leading to the transformation of the region into a 'wealthy zone'. This situation is described in the second part of Chapter Three.

Chapter Four further explores the nature of villagers' integration into the state. It is argued that the level of 'progress' that the Sumber Jaya and Way Tenong regions have achieved is largely the result of villagers' efforts to bring the state to the village as a strategy to tap state resources. The chapter outlines villagers' engagements with the state within the context of national politics, rural development and village administration.

Chapter Five illustrates the ways in which local people defend smallholder farming by resisting attempts by forestry authorities to exact greater control over land and people. Having been in conflict with forestry authorities for decades, after *reformasi* some of the villagers in the region agreed to engage in a new kind of relationship with forestry authorities. Collaboration between government and 'community' in 'sustainable natural resources management' is perceived to be the official goal of the new relationship. In practice, however, the desires of both parties are not easily reconciled and the struggle over control of land and resources continues.

Chapter Six outlines the history of the formation of Gunung Terang as an administrative village and focuses on the village's organisation: its administration; leadership; and sub-divisions. The chapter considers this village in the context of internal community affairs as well as within the framework of wider village relations. It is argued that the village's collective strategy is to mediate official relations between people and the state as well as within the community.

Chapters Seven and Eight, which focus on the village economy, are devoted to examining the persistence of smallholders. They explore the flexibility of smallholding agriculture, beginning in Chapter Seven with a discussion of socioeconomic differentiation among villagers. This discussion is then followed in Chapter Eight by a closer look at the persistence, modification, and alteration in farming systems. by a closer look at the persistence, modification, and alteration in farming systems. The chapter concludes with a discussion of the dynamics of the social organisation of smallholder agricultural production (land, labour and capital).

Chapter Nine summarises the trends discussed in the previous chapters.

2. Lampung in the Twentieth Century: The Making of 'Little Java'

Like many areas in Indonesia's outer islands, up until the mid-1900s Lampung was a sparsely populated, virtually 'empty' land. Lampung was known as the world's leading pepper producer. By the end of the twentieth century, Lampung produced surpluses of rice and other agricultural commodities along with pepper. The province was classified as extremely poor and, like Java, had an overpopulation problem. This chapter examines the driving forces behind these changes. Colonial and post-colonial government development initiatives such as land alienation and consequent forestry and other plantation establishment, decentralisation of administration, regional development, transmigration, and spontaneous migration are all identified as key forces that have transformed Lampung during this time.

Colonising 'The Empty Land'

In the past, Lampung was known as the world's 'pepper basket'. When neighbouring Banten developed into a flourishing international trading port in the seventeenth and eighteenth centuries, its main export commodity was pepper, which came largely from Lampung. From Lampung's coasts and navigable rivers, boats loaded with sacks of pepper regularly sailed across the Sunda Strait to Banten. Maintaining steady supplies of pepper from Lampung to Banten and later to Batavia was one of the top agendas of the Dutch trading company (VOC) and the subsequent colonial administration (Kingston 1987; Sevin 1989). The annual pepper production in Lampung steadily increased from 400 to 600 tonnes in the early 1800s, and to 2,000 tonnes in the 1880s. It increased even further to 4,000 tonnes in 1890, 10,000 tonnes at the turn of the twentieth century, and to 45,000 tonnes in the 1930s, making Lampung the source of 60 per cent of the world's pepper production (Bulbeck et al. 1998: 68).

The monopoly over pepper exports from Lampung has long been a source of rivalry between regional and international power centres. The Banten Sultanate controlled pepper supplies in most of the southern part of Lampung, in fierce competition with the Palembang Sultanate at Lampung's northern tip. Both fell under Dutch control in the first half of the nineteenth century (Kingston 1987; Sevin 1989). Bugis and Malay seamen were heavily involved in this trade, either by offering a higher price or by simply pirating the shipments. The British, who were denied access to pepper from Batavia and Banten, controlled the western part of Lampung (then part of Bengkulu Residency) from the 1680s until the

British transferred Bengkulu to the Dutch in exchange for Singapore in 1825 (Bastin 1965). In the second half of the nineteenth century, the Dutch were able to unify and place the southern part of Sumatra (Lampung, Palembang and Bengkulu) under their control. To ensure the flow of profit, the Dutch re-oriented trading routes in addition to forcing delivery of pepper and coffee to their warehouses for a set low price. Batavia was designated as an obligatory transit point for all export commodities, cutting the trading networks between southern Sumatra and Singapore (Sevin 1989).

In addition to pepper, coffee gained in importance as an export crop from Lampung during the nineteenth century. Coffee cultivation began at the beginning of the nineteenth century (Sevin 1989: 45) and became a lucrative cash crop in Lampung by the century's end. Unlike in Java, where cash crop booms were largely the result of the infamous cultivation system (*cultuurstelsel*) imposed by the Dutch, in Lampung both pepper and coffee were cultivated by traditional smallholders (Bulbeck et al. 1998). These smallholder farmers 'supplemented shifting cultivation with cash crops' — an ancient farming method said to be typical of upland southern Sumatra (Pelzer 1945: 24–6). Coffee or pepper supplemented the traditional crop (such as upland rice) and the gardens were not permanent:

> The lifetime of coffee bushes on ladangs [open swidden fields] is only from 3 to 5 years once they have started to yield berries…. The lifetime of a pepper garden is considerably longer, perhaps 15 years…. [S]hifting cultivators plant coffee bushes 1.5 to 2 metres apart in the midst of upland rice fields during the first year that they occupy a ladang. In the following year the bushes are still small enough to permit the growing of a rice crop among them. In the third year a new ladang is made and planted with rice and coffee, while a coffee harvest is gathered from the first ladang. In the fourth year the first ladang produces an excellent crop of coffee. In the fifth year the first ladang yields its third coffee harvest and the second its first coffee harvest, while rice and coffee are planted in a third ladang. In the seventh year the cultivator may have as many as four ladangs, the first producing its last coffee crop before it is abandoned because of the declining yields, the second yielding its third coffee harvest, the third just entering the bearing stage, while the fourth ladang supplies the shifting cultivator with rice grown among young coffee bushes (ibid.: 25–6).

Such a method of cultivation was strikingly different from other methods of cash crop cultivation practised up until the twentieth century by other farmers in nearby regions. Hevea rubber in Sumatra and Borneo and benzoin in North Sumatra were produced in permanent gardens (Pelzer 1945). In Java, under forced cultivation, cash crops such as coffee and sugar were produced using (modern) intensive farming methods. In the case of pepper, Chinese migrants in a few areas of the Riau Archipelago, Malay Peninsula, Siam, Cambodia, and Brunei introduced intensive pepper cultivation where hardwood — instead of traditional *chinkareen* (also known as *dadap*) trees — was used to support the pepper vines (Bulbeck et al. 1998: 144–9). The ground was turned for clean-weeding twice a year and fertiliser was applied (ibid.). Yield in these fields was much higher (over 2,000 pounds an acre) compared with the traditional system in Sumatra. By contrast, in Bengkulu, annual yield per acre was just 310 pounds. The weakness of this 'Chinese method' compared with the traditional system was its inability to withstand price variations due to its high labour input and upfront cash outlay. This eventually led to its abandonment (ibid.: 155–6). Due to its low inputs, yields from traditional swidden agriculture were relatively high in terms of return to labour (Dove 1986), and could withstand significant export price variations.

Apart from pepper and coffee, forest products were also important export commodities from Lampung. Rattan, elephant tusks, rhinoceros horns, swallows' nests, rubber, and damar resin were exported to Batavia and Singapore during the mid-1800s (Sevin 1989: 45). On the other hand, rice was regularly imported from Java. Unlike neighbouring Java, Bali, and highland Palembang, the indigenous people of Lampung were not familiar or interested in constructing large irrigation networks. Wet rice fields were limited to the banks of streams and rivers. Swidden rice fields in this region provided the largest source of rice by far (Kingston 1987; Sevin 1989).

Smallholder production of rice, pepper and coffee using traditional farming systems was possible largely because of the low population density in Lampung as a consequence of land abundance and labour shortages. The Lampung population was merely 104,200 people in 1845, and though that number had nearly doubled 60 years later, it still gave an average density of less than five persons per square kilometre (Sevin 1989: 47). In contrast, over the same period, Java's population more than tripled from 9.3 million to 30.1 million resulting in an average density exceeding 200 persons per square kilometre. Compared to Java, Lampung at the beginning of the twentieth century was indeed an 'empty land'.

Besides its low population density, Lampung's population at the turn of the twentieth century was also unequally distributed (Sevin 1989: 47–8). The mountain range of Bukit Barisan, apart from Balik Bukit and Belalau in the northwest and the swampy plains and estuaries of large rivers such as Tulang Bawang and Seputih in the northeast, was largely devoid of human settlements. Villages and small towns were scattered along the south and west coasts and on the banks of navigable inland rivers. In the mid-1800s, 21,270 people inhabited the Krui coast in the west, 12,000 people lived in Semangka Bay in the southwest, and 16,690 people occupied Lampung Bay in the southeast. By the turn of the twentieth century, ports on these coasts were developed into small towns and commercial centres: Krui on the west coast; Kota Agung on Semangka Bay; and Teluk Betung on Lampung Bay. Teluk Betung, with 4,500 inhabitants, was the largest town.[1]

Farther inland, the banks of the Way Sekampung River and Labuan Maringgai further downstream were home to some 10,600 people. Way Seputih River and its tributaries (Way Pegadungan, Way Sukadana, Way Pengubuan, and Gunung Batin) in the centre had 14,118 inhabitants, and Way Tulang Bawang and its tributaries on the north (Way Abung, Way Sungkai, Way Kanan, and Way Besai) were occupied by 29,450 people. Small towns located on the banks of large rivers included Menggala — then the largest inland town — on the Way Tulang Bawang River, Gunung Sugih, and Terbanggi on the Way Seputih River and Sukadana and Labuan Maringgai downstream of the Way Sekampung River.

Around those small towns, indigenous Lampung houses were grouped into traditional villages. The villages, located along the river banks and separated by a few kilometres, had a few hundred and sometimes up to 1,000 inhabitants (Sevin 1989: 47). Between the villages was uninhabited land where temporary hamlets could be encountered. These hamlets were created near the newly opened swidden fields (*ladang*) far from the villages (Utomo 1975; Kingston 1987).

The indigenous Lampung divided themselves into three large sub-groups — Pesisir, Abung, and Pubian. The Pesisir occupied Lampung's west and south coasts, Abung dominated inland rivers in the centre and north, while Pubian settled in the smallest area in the centre and south. Other smaller sub-groups, such as Menggala on the Way Tulang Bawang River and Meninting and Melinting in Maringgai downstream of the Way Sekampung River, are the result of the mixing of the main Lampung groups with outsiders. Menggala is the fusion of Pubian, Malay and Bugis, while Meninting and Melinting groups are comprised of Pesisir and Pubian people with others from Banten, Sunda, Java and Bugis (Sevin 1989: 49–69).

---

1   The population of Teluk Betung and the surrounding Lampung Bay and Semangka Bay was severely affected by the powerful Krakatau eruption in 1883. The huge tidal waves caused by the eruption wiped out villages and killed thousands of people.

Map 2-1: Lampung Province.

Dutch scholars and official reports provide a brief account of social organisation amongst the indigenous Lampung population (Utomo 1975; Kingston 1987; Sevin 1989). Two terms — *marga* and *buay* — were commonly used to describe the population. *Marga*, representing the largest socio-political unit, consisted of a number of genealogically related villages (Pekon, Tiuh, Anek, Dusun, or Kampung) and emphasised territoriality. *Buay* put more emphasis on genealogical ties. *Buay* were divided into smaller patrilineal groupings known as *suku* (lineage), which were further divided into *cangkai* (sub-lineage) and then *nuwo* (houses). The relationship between *marga* and *buay* varied. In some areas a *buay* was a *marga*. In other areas, several *buay* comprised a *marga* or a *buay* consisted of more than one *marga*. Several *marga* and/or *buay* often formed federations. At the village level, *suku* leaders (*penyimbang*) met in assembly (*proatin*) to govern village affairs. Similarly at the *marga* level, the assembly consisted of village-level leaders (*penyimbang*) of the *marga*. The indigenous people of Lampung observed male primogeniture in the inheritance of title and property, and authority was based on seniority. The oldest village from which other villages had split was the seat of the *marga*. A *suku* leader was the eldest male descendant of the founder of a *suku* and this status was granted by the assembly followed by a title granting feast ceremony (*pepadun*).[2]

Lampung social organisation was considerably influenced by the indigenous population of Lampung which was predominantly Muslim. The Banten Sultanates had much more influence on the external affairs of the *marga* (Sevin 1989: 51–9). The four *marga* (*marga pak*) around Mount Rajabasa in Lampung Bay, for example, were ruled by a lord (*ratu*) who was under the command of the Banten Sultanate. Similarly, Ratu Melinting ruled the Sekampung valley in Banten's name. Elsewhere, among several adjacent *buay* or *marga,* the Banten Sultanate granted noble titles. A *jenjem* (political representative) was appointed to supervise a number of *paksi* and *bandar* (local representatives) who were selected among the *buay* or *marga* chiefs. The power of Banten over Lampung, however, was limited to ensuring the monopoly over pepper. The granting of titles signalled a person's nobility, denoting high social standing but without political authority. This was another element of indigenous Lampung social organisation understood to have been introduced from Banten.

In the mid-1800s, after decades of military campaigns, the Dutch were able to overcome the indigenous rebellions. By this time the Dutch had shifted from a monopoly over pepper toward control over the land and its people (Kingston 1987). Following their conquest, the Dutch imposed a formal

---

2   The indigenous Lampung population now identify themselves either as belonging to Lampung Pepadun or Lampung Pesisir (Hadikusuma 1989). The Pepadun inhabited inland Lampung and the status of village leader (*penyimbang*) is granted to them by the assembly only if the incumbent is able to perform the expensive ritual and feast of *cakak padun*. For the Pesisir or Peminggir, inhabiting Bukit Barisan mountain range and Lampung's southern and western coasts, the *penyimbang* status is hereditary.

administration and Lampung was divided into five (later seven) sub-divisions (*onderafdeeling*), each of which had a Dutch officer (*controleur*) supervising an appointed non-indigenous official (*demang*) from either Java or Palembang who dealt directly with individual villages. The new structure ignored the higher traditional political unit of *marga*. Further, the Dutch declared that the vast uncultivated lands between villages, which had traditionally been *marga* lands, had become part of the state domain.[3] Some of these former *marga* lands were then granted under long lease to private estate plantations, with many more designated as forestry zones (*boschwezen*). In 1928, the Dutch recognised the *marga* as a political unit but modified many of its principles to meet government ends. Some of these modifications included collecting taxes and recruiting corvée labour for the construction of roads.[4] *Marga* control over land outside individual villages was never fully given back.

In the first half of the twentieth century, a major aim of the Dutch administration was to develop colonies of Javanese in Lampung. A railway line was constructed from Teluk Betung to Palembang, alongside which enclaves of 'little Java' would be created. This was supposed to be followed by the development of a plantation belt (*cultuurgebied*) (Kingston 1987). The Dutch started to create agricultural colonies of Javanese in Lampung in 1905. The plains on the south side of Lampung were selected as the primary sites: the first developed colony was Gedong Tataan; the second Wonosobo; and the third Sukadana. By 1941, 174,000 people, mostly from the overcrowded areas in central Java (Kedu, Banyumas, Pekalongan, Yogyakarta, Surakarta and Jepara/Rembang) and a smaller number from east Java (Kediri, Surabaya, Madiun and Malang), had been settled in the new colonies (Sevin 1989). Sukadana was the largest settlement with 90,000 inhabitants and its main village of Metro was turned into a town in the 1930s (Hardjono 1977). Each colonist received 0.3 hectares of dry field (*tegalan*) and 0.7 hectares of irrigated rice field (*sawah*). Under the colonisation scheme, the colonists brought the Javanese style of lower-level village structure (*desa*) and sub-district (*kecamatan*) administration with them to Lampung.

Opposition to this colonisation scheme came initially from plantation companies in Java and North Sumatra who claimed that they had experienced difficulties in recruiting labourers because they would rather participate in government-sponsored migration than become cheap 'coolies'. The Great Depression of the 1930s also forced plantation companies to stop recruiting and to reduce

---

3 Previously all Lampung land was divided among its *marga*. On average, a Lampung *marga* had 5000 inhabitants, less than 10 villages, and occupied 500 square kilometres (van de Zwaal 1936, cited in Utomo 1975: 52). Up until the 1950s, there were 87 *marga* in Lampung (Hadikusuma 1989: 189–94) of which 81 were made up of indigenous Lampung. Six *marga* comprised migrants from South Sumatra, four *marga* from Semendo, and one each from Ogan and Mesuji/Pegagan.

4 Kingston (1987) suggests that the abuse of power by the new government-selected *marga* chiefs gave rise to popular protests organised by the Commite Tani Lampung (Lampung's peasant committee) in the 1930s.

the number of their workers. This revived the colonisation projects and led to the creation of the last and largest colonial agricultural resettlement zone of Sukadana. Argoguruh Weir was constructed to channel water from the Way Sekampung River to the zone's irrigation schemes. The colonisation project was stopped with Japan's invasion in 1942, the subsequent World War Two, and Indonesia's revolution for independence.

Government-sponsored migration from Java to Lampung was re-started in the early 1950s and the program was renamed transmigration. Under a project called Corp Tjadangan Nasional (CTN), which was later renamed Biro Rekonstruksi National (BRN), former soldiers and militias from various part of Java were moved to Lampung. In the early 1950s, small groups of these veterans were given cleared land and were expected to clear further areas to attract more Javanese. About 25,000 people migrated to Lampung, 60 per cent of whom still remained in the 1960s (Benoit 1989: 107). Unlike the Dutch colonisation and later post-colonial Indonesian transmigration projects that concentrated on areas in the eastern lowlands of Lampung, the BRN also placed Sundanese and Javanese migrants in the western highlands. Lowland Palas in the south and Jabung in the centre were allocated to the BRN transmigrants. In the eastern foothills of the Bukit Barisan mountain range, four sites were selected: Pulau Panggung in the south; Kalirejo in the centre; and Tanjung Raya and Sumber Jaya in the north.[5]

Subsequent transmigration programs largely followed Dutch patterns. The central and northern plains were designated as the transmigration receiving areas. From the mid-1950s until the end of the 1970s, the plains around Sukadana, Gunung Sugih, and Kota Bumi were transformed into transmigration receiving areas. The World Bank was the main sponsor of the post-colonial transmigration program. Between 1950 and 1969, 100,000 hectares of lands were allocated to 200,000 transmigrants. The number of transmigrants who settled on 53,000 hectares of land fell to 50,000 between 1969 and 1974, and finally dropped to 11,000 people between 1974 and 1979 (Pain 1989: 293–4). Unlike colonisation schemes, post-colonial transmigration programs did not always allot an irrigated wet field (*sawah*) to each transmigrant, and many transmigrants only received a dry field plot (*tegalan*). The total area of land given to transmigrants was somewhat larger, being 2 hectares or more.

By the end of the 1970s, Lampung ceased to be the destination of Javanese transmigrants. The local transmigration program (*transmigrasi lokal* or *translok*) was redesigned to remove forest squatters from government-designated forestry zones and to 'develop' the isolated, sparsely populated northeast regions of

---

5   Earlier and subsequent development of one of these sites, Sumber Jaya, is the subject of the following chapters.

Lampung. From 1979 to 1986, over a quarter of a million people were forced to move from the southern and central forestry zones to the plains and swamps between Menggala, Mesuji, and Blambangan Umpu. Several areas of 100,000 hectares each were cleared for this purpose.

As a consequence of the influx of migrants, the proportion representing the indigenous population fell dramatically (Benoit 1989: 143–5) — from 70 per cent of the population in 1920 to less than 15 per cent in the mid-1980s. In the mid-1980s, nearly 70 per cent of the Lampung population was Javanese, with the Sundanese being slightly more than 10 per cent, and migrants from South Sumatra being less than 10 per cent.[6]

Table 2-1 shows the total non-indigenous Lampung population in the second half of the twentieth century. The number of spontaneous migrants or independent settlers and their descendants who migrated to Lampung without government assistance was much greater than the number of government sponsored migrants and their descendants. Following their friends and relatives who had been sponsored to migrate to Lampung, other Javanese and Sundanese sold their possessions in Java to buy land in Lampung. Those who did not have enough money settled to work as labourers and/or sharecroppers for the earlier migrants and indigenous Lampung smallholders (Utomo 1975; Levang 1989).

Table 2-1: Population of Lampung Province, 1930–86.

	1930	1961	1971	1980	1986
Population outside Bandar Lampung					
Natives of Lampung	218,000	360,000	458,000	556,000	661,000
Spontaneous migrants		498,000	577,000	1,057,000	530,000
Descendants of spontaneous migrants	123,000	184,000	804,000	1,652,000	2,340,000
Transmigrants (excluding translok)		375,000	107,000	135,000	199,000
Descendants of transmigrants	35,000	55,000	513,000	755,000	1,002,000
SUB-TOTAL	376,000	1,472,000	2,459,000	4,155,000	4,732,000
TOTAL (including Bandar Lampung)	406,000	1,667,000	2,775,000	4,627,000	5,250,000

Source: Benoit 1989: 130,168.

---

6   Other migrant ethnic groups in Lampung are Chinese, Minangkabau, Bugis, Balinese, Batak, and Madurese.

Labour migration to Lampung is not a recent phenomenon. For centuries, groups of labourers from Banten had come to Lampung to handpick the pepper corns and coffee cherries. These seasonal migrations involved 30,000–40,000 people per year at the turn of the twentieth century. Although they usually returned to Banten, some settled in Lampung (Kingston 1987). Three large groups — the Mesuji, Ogan and Semendo (Benoit 1989) — migrated to Lampung at the end of the nineteenth century and the beginning of the twentieth century from the neighbouring province of South Sumatra (Palembang). Of these three: the Mesuji moved from the lowland Palembang border with Lampung to northeast Lampung; the Ogan moved to the northern Lampung plains between Kota Bumi and Bukit Kemuning; and the Semendo moved from their homeland in highland Palembang to the hilly and mountainous highlands of Lampung such as Kasui and Way Tenong in the northwest and Pulau Panggung in the southwest. The Semendo, the largest of the three groups, cleared the jungle and transformed it into coffee gardens. They settled in villages or hamlets near streams where, as in their homeland, they could establish wet rice fields. The Semendo and Ogan migrants often employed and sold their coffee gardens to incoming Javanese migrants.

Thus, the transformation of Lampung in the twentieth century was a result of complex factors. Some of these relate to the influx of migrants and the subsequent opening up of Lampung land. While some forces have attempted to control or limit the movement of these migrants, others have worked to benefit from their arrival.

Besides colonial and post-colonial government-sponsored transmigration, infrastructure development and administration decentralisation also stimulated spontaneous settlers to Lampung and, as a consequence, the opening up of its land. The Dutch constructed a railway from Teluk Betung at Lampung Bay to Palembang in the 1920s. It is along this railway line that most populated areas and economic centres are located. Road networks were continuously built to connect remote areas to these population and economic centres. Initially constructed by corvée labour during the Dutch occupation, the construction of these roads became the main post-colonial development agenda, especially after 1970 when transmigration programs were integrated into regional development. As a result, commerce was boosted, more lands were cleared and cultivated, and more spontaneous settlers moved in.

Colonial and post-colonial administration decentralisation and land alienation played an important role in the transformation of Lampung. In the mid-1800s, the Dutch dismantled Lampung's traditional government of *marga*, first by imposing a modern administration with each district (*onderafdeeling*) headed by a Dutch officer assisted by several non-indigenous workers (*demang*) to work directly with officially selected village heads (Kingston 1987). When the Dutch declared the vast areas of uncultivated land between villages as state property, the indigenous Lampung lost important reserve lands vital to the continuation of their traditional agricultural production systems. The indigenous people were soon unable to resist migrant influxes from the north (Semendo, Ogan and Mesuji) that had moved in to occupy former *marga* lands with the Dutch officer's consent. A large portion of the former *marga* lands were either granted on long-term leases to plantation companies or designated as forestry zones.

The Dutch created a different style of local-level administration, one for Javanese settlers and one for the indigenous population. From 1928, the villages (Pekon, Kampung, Dusun, and Tiuh) indigenous to Lampung were organised under *marga* headed by chiefs (*pesirah*), who were selected by the government among the village leader nominees (Kingston 1987). Javanese transmigrant villages were organised into sub-districts, headed by a government officer (Hardjono 1977). Several sub-districts formed one municipality (*kewedanan*) which was led by a district administrator (*wedana*). Lampung still largely retained this dual system until the end of the 1970s, well into the post-colonial era. A modification was introduced for the indigenous Lampung population in the 1950s (Utomo 1975) by merging several *marga* into a *negeri* (sub-district) led by a chief (*kepala negeri*) who was selected among the village leaders. A *dewan negeri* (sub-district council) was also formed as the council of indigenous elders. As more Javanese migrants settled in Lampung, former *marga* lands were converted into Javanese settlers' villages. The takeover of forests and bush land was usually marked by a token payment (*ulasan*) by settlers to the head of the *marga* or *negeri* (Utomo 1975). Since the Dutch forbade the indigenous population to sell their land to the Javanese migrants, *ulasan* was only paid as compensation for the loss of cultivated plants. In the late 1970s, the merger of several *marga* (*negeri*) was abolished and all of Lampung then adopted the Javanese style of administration.

The last three decades of decentralisation of administration resulted in the continuous creation of new districts, sub-districts, and villages. Lampung had four districts, 60 sub-districts, and 1,164 villages in 1972; four districts, 77 sub-districts, and 1,941 villages in 1991; and 10 districts, 162 sub-districts, and 2,099 villages in 2001 (BPS Lampung 1972 and 2002). The direct impact of the creation of these government administrative units (*pemekaran wilayah*) has been more rural development programs such as roads, schools, clinics and agricultural extension.

Since the early days of government-sponsored migration from Java to Lampung, the estate plantation sector has benefited from abundant and cheap labour. The opening of colonisation sites by the Dutch and the influx of Javanese agricultural colonists up until the 1940s were soon followed by the granting of long leases on nearby land to plantation companies. Estate plantations of coffee, rubber and oil palm were created close to the transmigration settlements. In the 1930s, there were 34 plantations with sizes between 2,000 to 5,000 hectares that employed seasonal labourers from the transmigration settlements (Kingston 1987). After Indonesian Independence, many of these plantations were nationalised and were placed under the control of PT Perkebunan Nusantara, a state-owned plantation company.

Following the opening of new transmigration settlements in central and north Lampung, post-colonial era state-owned and private companies were granted land leases for new estate plantations. The size of many of these new estate plantations was much larger than those of the colonial period. Some private companies had 20,000 hectares of land, while others controlled as 'little' as 40 hectares. Coffee diminished in importance and was no longer an estate plantation commodity. In addition to rubber, oil palm and sugar, other crops such as cassava, coconut, pineapple, and banana also became estate plantation crops. In 1969, estate plantations controlled 21,000 hectares of Lampung land, and by 1985 that number had risen to 133,000 hectares (Pain 1989: 347). In the 1990s, Lampung's eastern swamps and coasts were gradually transformed into brackish shrimp ponds. In the late 1990s, Lampung was the home of two of the nation's largest shrimp industries.

As elsewhere in Indonesia, forestry policies in Lampung have for decades been designed and implemented to exclude 'undesirable' people and their associated land uses. The Dutch started the process first by confiscating indigenous *marga* lands and declaring them to be part of the state domain in the mid-1800s. Subsequent land alienation was carried out through the designation of forestry zones. Between 1922 and 1943 nearly a million hectares of Lampung lands were designated as forestry zones. The indigenous population was prohibited from both harvesting forest products and clearing the land for farming. The Dutch controlled the harvesting of forest products and gave the *marga* only a small share of the income (Kingston 1987; Utomo 1975). A plan was drawn up for forestry plantations of teak such as those in parts of Java, a lucrative source of income for the Dutch. Labourers from Java were brought to Lampung and some hundreds of hectares of forestry zoned land between Gedong Tataan and Tegineneng were planted with teak. However, the Japanese invasion in 1942–3 prevented further materialisation of the plan (Utomo 1975). While strictly prohibiting the indigenous people from gathering forest products, as well as clearing the land within gazetted forestry zones, the Dutch were more permissive toward the Javanese transmigrants whose allotted settlements and fields were already fully utilised. These transmigrants were allowed to clear forest land and attempt to extend their agricultural settlements (Kingston 1987). Sections of the colonisation zones between Gunung Sugih and Sukadana were forestry zoned lands that were converted into settlements.

Immediately after independence, logging became the main type of forestry work in Lampung. Prior to their designation as national parks in the 1980s, sections of Way Kambas (in East Lampung) and Bukit Barisan Selatan (in West Lampung) were also granted for logging concessions. In addition to former forestry zoned lands, thousands of hectares of the remaining *marga* lands were granted to logging companies. Once this was done, these former *adat* lands were officially classified as state forestry zones, which then legally became the property of the state. Meanwhile, some of the forestry zones were converted into transmigration and spontaneous settlements, such as the development of post-colonial pioneer transmigration settlements on the edges and within the boundaries of forestry zoned land such as in lowland Palas and Gunung Balak and in highland Pulau Panggung and Sumber Jaya. Other former forestry zones and logging concessions were converted into estate plantations. By the early 1990s Lampung had no more production forest. About 175,000 hectares of former logging concession areas are now officially under the control of the state forestry company (PT Perhutani) or industrial forestry plantations (*hutan tanaman industri*).

In the last three decades, the designation of state forestry zones, reforestation, and the eviction of 'forest squatters' have become key forestry policies. About 1.2 million hectares of land, over 30 per cent of Lampung land, consisting of mostly former Dutch forestry zones plus post-colonial logging concession areas have been reclassified as state forestry zones (*kawasan hutan negara*). From the late 1970s to the early 1990s, at least a quarter of a million people were forced to vacate the protection forest zones (*kawasan hutan lindung*) in the upper part of the watersheds to join local transmigration programs. This was then followed by the planting of exotic trees such as rosewood (*Dalbergia* sp.) and calliandra (*Calliandra calothyrsus*) on the abandoned smallholders' fields and settlements. But the plan to transform these forestry zones into forestry plantations was never fully implemented; the reforestation trees died, were overgrown by bush, or were removed and the areas were transformed back into smallholder farmers' fields. The appropriation and reappropriation of land in forestry zones has been a constant feature of the interaction between local people and forestry authorities in Lampung.

The forestry authorities used conservation of watersheds as the reason to justify the imposition of repressive forestry policies and the selection of forestry zones in 'water catchment areas' of Lampung's main rivers. Large dams were constructed on the upper reaches of these rivers to feed water for the irrigation canals downstream and/or for hydroelectric power (DPU 1995). The Way Jepara dam, located near Gunung Balak in East Lampung, was constructed between 1975 and 1978 and is designed to irrigate 6,651 hectares rice fields in the Way Jepara region.

Located at the upper Way Rarem River near Kotabumi, the Way Rarem Dam was constructed from 1980 to 1984 to irrigate 22,000 hectares of rice fields. At the upper part of the Way Sekampung River, near Pulau Panggung and Talang Padang in the Tanggamus Highlands, the Batu Tegi Dam was constructed from 1995 to 2003. The Batu Tegi Dam was designed to produce 24 megawatts of electricity and to supply water for 90,000 hectares of irrigated rice fields on the eastern and central Lampung plains. The construction of the Way Besai Dam, located in Sumber Jaya in the West Lampung Highlands and designed to produce 90 megawatts of electricity, was started in 1994 and completed in 2001. Financial support for the construction of these dams came primarily from the World Bank and the government of Japan. The eviction of thousands of migrant smallholder families in Gunung Balak since the early 1980s was described as a necessity to ensure a steady supply of water for the Way Jepara Dam. The construction of the Batu Tegi Dam was preceded in the early 1990s by a massive demolition of migrant smallholders' houses and coffee gardens which were replaced by *sonokeling* (rosewood) plantings in Pulau Panggung, Wonosobo, and Sumber Jaya. More recently, similar attempts have been conducted on the upper parts of the Way Tulang Bawang River (such as Tanjung Raja, Bukit Kemuning, and Sumber Jaya).

Land of Hope, Land of Despair

While criticising some aspects of the Dutch transmigration projects in Lampung up until the 1940s, Karl Pelzer (1945) also praised the projects for their potential to increase population redistribution and intensive agricultural production. A decade later, instead of intensive agriculture, J.F. Wertheim (1959) encountered vast areas of *alang alang* grasslands (*Imperata* sp.) replacing the forest cover he had seen in the 1930s. Instead of well-planned agriculture settlements, he found the spread of spontaneous settlements — a result of the saturated early transmigration sites — whose populations were in desperate need of government assistance if their livelihood was to improve. Comparing the conditions of pioneer agriculture settlements of Javanese migrants in Lampung in the 1930s and the late 1950s, Wertheim indicated that in the future this 'land of hope' could turn into a 'land of despair'.

Kampto Utomo[7] gave a detailed account of village social organisation and the modes of ecological adaptation of the spontaneous settlers in Lampung in the mid-1950s (Utomo 1975). Between 1950 and 1957, spontaneous migrants from nearby old transmigration sites (for example Gedong Tataan and Pringsewu) and

7   He later changed his name to Sayogyo. Professor Sayogyo is a well-known Indonesian rural sociologist who focuses on rural development and poverty alleviation.

directly from Java opened new agricultural settlements in the Way Sekampung area.[8] Between 35,000 and 40,000 inhabitants created 18 new Javanese villages. The settlers were recruited by a number of the chiefs of the clearing (*kepala tebang*) who sought permission and paid *ulasan* compensation to the indigenous Lampung sub-district (*negeri)* head to clear the forest. The forest was cleared collectively and male settlers each received a farming field and housing lot. More migrants came and more forests were cleared. Under the leadership of the chief of clearings (*kepala tebang*), several hamlets formed administrative villages, usually with one of the *kepala tebang* as the village head. Government assistance was absent, so roads, markets, schools and clinics were constructed through community work. The village administration received a tax (*janggolan*) either in cash or in kind (such as rice) from the villagers. Later the new villages were organised into a sub-district led by a government officer. However, the much needed government assistance (for the construction of roads, irrigation networks, and agricultural extension) was still absent.

Rather than practising the intensive agriculture that they knew so well from Java, these spontaneous migrants adopted the indigenous people's method of shifting cultivation, but at its worst form in an ecological sense (Utomo 1979). As described by Pelzer (1945), the shifting cultivation practised by the indigenous population was supplemented by cash crops. This usually involved one or two crops of upland rice on an area of newly cleared forest, after which the field was planted with coffee and/or pepper before it was left fallow. The Javanese migrants prolonged the planting of annuals (rice and soybean) and left the field fallow for a short period so that the soil was rapidly exhausted and became infested with *alang alang* grass (Utomo 1979). The migrants managed to convert stream banks into flooded rice fields, but the irregular water supply and lack of labour limited production. A few of them tried to plant coffee, but low production and poor upkeep soon transformed their coffee gardens into *alang alang* fields. These migrants then abandoned their fields and searched for new forests to clear, starting a new cycle of conversion of forest cover into *alang alang*.

Concerns about the livelihood of the population in Lampung, with an emphasis on the distribution of population and agricultural production systems, continued to be raised by a French research team in the mid-1980s.[9] Many areas were already heavily populated and saturated, while a few zones called 'last frontiers' were scarcely populated (Pain 1989). Agricultural practices underwent profound changes. Patrice Levang's survey (1989) demonstrated that irrigated

---

8   This area forms the present day Kalirejo and Sukoharjo regions.
9   The research team came from the Office de la Recherche Scientifique et Technique d'Outre-Mer (ORSTOM).

rice fields, gardens, perennial cash crops and dry fields with annual food crops were the main farming systems of Lampung by 1980s. Often these systems were practised as a mixed farming system.

Pain (1989) divided Lampung into three forms of spatial organisation: 'the centres'; transition areas; and marginal zones. The 'centres' had populations of over 500 inhabitants, and in some areas over 1000 inhabitants, per square kilometre of cultivated land. These centres were rice-growing plains created by colonial and post-colonial transmigration programs. They included: Pringsewu, Metro, and Bandar Jaya and their surroundings; the piedmont parallel with the Bukit Barisan mountain range from southwest to northwest (from Talang Padang in the south, to Kalirejo and Sukoharjo in the centre, to Kota Bumi and Bukit Kemuning in the north); and large-scale estate plantations occupying the same areas. The Talang Padang area was dominated by irrigated rice fields and gardens; in Kalirejo and Sukohardjo, irrigated rice fields and gardens and dry fields were mixed; and in Kota Bumi and Bukit Kemuning, pepper and coffee were the predominant crops.

The transition areas were those on the peripheries of the rice-growing plains, the non-irrigated plains from Palas and Sidomulyo in the south to Sukadana, Gunung Sugih, and Padang Ratu in the centre, as well as the west and south coasts (Krui and Kalianda). The population density of these areas was 500 persons per square kilometre or less. Rain-fed rice fields were farmed on low-lying marsh land in these transition areas, but dry-land fields — planted with mixed or rotational annual food crops of maize, soybean, peanut, mung bean, and cassava — were becoming more dominant. When the soil deteriorated, cassava was the only crop that would grow, otherwise the land was taken over by *alang alang*. The exception to this pattern was the farming systems on the south coast of Kalianda and on Rajabasa Mountain. Here the coasts were dominated by village-scale cultivation of irrigated rice fields and coconut groves, while the adjacent hill slopes were transformed into perennial cash crop gardens. Prior to the mid-1980s, cloves inter-planted with coffee were the main crops that brought prosperity to these regions.

The marginal zones were sparsely populated areas and consisted of 'isolated areas' and 'enclaves' with a population density below 200 persons per square kilometre. The isolated areas were the last transmigration settlements in Central Lampung (Rumbia, Seputih Surabaya, and Seputih Mataram) and North Lampung (Panaragan, Way Abung, and army veteran transmigration sites near Kotabumi), together with the newly cleared mountain areas of Sumber Jaya, Kenali, Liwa, and their surroundings. In the last transmigration settlements, 'stagnant' poverty was the main feature. Due to poor soil and isolation only cassava could grow. Working as seasonal labourers on the nearby large estate plantations provided another source of income, but wages paid to men, women and children (who had typically dropped out of school) were low. The situation contrasted with mountain areas which were still being progressively cleared. Although also isolated due to the absence of road networks, such areas were endowed with fertile soil and higher and longer rainfall. Returns from forest clearing, a crop or two of upland rice, coffee gardens, and (in Liwa) vegetable fields (cabbage, potato, shallot, and chilli) were high. In these pioneer zones, Pain (1989: 341) notes that 'here and there wealthy zones have taken shape'.

Included in the enclaves of the marginal zones was the Krui region on the west coast, local transmigration clearings in the northern part of Lampung, swamps under reclamation on the east coast (Rawasragi in Palas in the south and between Way Tulang Bawang and Way Mesuji in the north), and various forest reserves. As on the Kalianda coast, village-scale irrigation of rice fields and coconut groves could be seen in the surrounding village settlements of the indigenous Lampung people on the Krui coast. But unlike in Kalianda, where the hills returned to coffee after the demise of the clove gardens, the hills were returned to damar (*shorea javanica*) resin gardens in Krui. Local transmigration sites in Mesuji and Tulang Bawang were in their early stages, irrigation canals were under construction and the transmigrants were struggling to survive, supporting themselves by growing food crops. Another irrigation canal built in the early 1980s was the Rawasragi in Palas in the south which was allocated to transmigrants and spontaneous migrants who had been in the area since the 1950s, and had subsisted for decades by farming dry and flooded rice fields.

One of the features of Lampung depicted by Levang (1989) and Pain (1989) is the marked heterogeneity of the livelihoods of the rural Lampung population. They described zones of wealthy villages adjacent to zones of poor villages, and within the villages, wealthy families neighbouring poor families.[10] The initial increase in population, shrinking land holdings and decreasing production per capita, with subsequent decreases in household incomes, later characterised villages in both wealthy and poor zones. In the poor villages, the problems worsened much more rapidly.

Villages with families whose incomes were above subsistence level were found in the zones endowed with fertile soil along with gardens, irrigated rice fields, and mixed gardens and dry fields. From south to north, gardens dominate in the foothills of Bukit Barisan mountain range and its adjacent plains (Talang Padang, Kota Bumi, Bukit Kemuning, Sumber Jaya, and Balik Bukit), and the coasts (Kalianda, Kota Agung, and Krui). The fertile plain around Sukadana (Sukadana, Pugung Raharjo, Labuan Maringgai, and Jabung) is also dominated

---

10   A wealthy village has a relatively large proportion of families whose income enables them to afford other things beyond basic subsistence needs (such as food). Villages whose inhabitants were mostly families with income below subsistence level were defined as poor villages.

by gardens and is often mixed with productive dry fields. The transformation of swidden fields into gardens was brought about by the increasing amount of labour available for weeding, regeneration and harvesting thanks to the influx of spontaneous migrants. Labour arrangements (daily wages, contracts, and sharecropping) enabled the migrants to accumulate savings and to buy their own gardens. One hectare or two of garden, the size that can be managed by an average family, provided surplus family income. However, gardens were constantly being divided through the bequeathing of land to children, which led to a decrease in the average size of landholdings.

As the number of migrant labourers grew due to natural population growth as well as new arrivals, the area of land available for sale declined. When there was no more empty land nearby to clear for expansion, a stratum of landless labourers emerged. Villages in zones with irrigated rice fields were also home to families with incomes above subsistence level in the mid-1980s. Irrigation canals, villages, and road networks in these zones (notably Wonosobo, Pringsewu, Gedong Tataan, Metro, and Bandar Jaya) were built during colonial and post-colonial transmigration programs. During the colonial period, the allotment that a transmigrant family received was 0.7 hectares of rice fields. Later, under post-colonial transmigration, the allotment was increased to 1 or 2 hectares. Regular water supply, fertilisation, agricultural extension and a greater labour input significantly increased the production of rice in these zones. The first and second generations of transmigrants lived a better life than the one they had in Java, but this period of prosperity was short-lived. The division of land through inheritance, a sharp increase in land prices, and a growing population soon reduced per capita production and income. For the first and second generations of transmigrants, the variation of income among families was small, but after three or four generations the gap increased. A few rich families benefited from rice re-selling, shops, and huller machines, while the proportion of landless and near landless families has significantly grown. As a response to this pressure, the population in these zones employed various strategies such as off-farm and non-farm employment, cottage industries, and migration within and outside the province.

Poor villages with dry fields as the main farming system were scattered on the plains with poor soil. These were the last transmigration settlements that had no irrigation canals. These areas were from Seputih Banyak to Seputih Surabaya in the centre, which Joan Hardjono (1977) termed 'cassava villages', to the local transmigration sites near Menggala in the north, taking in the nearby villages created by subsequent spontaneous migrants. Annual food crops such as upland rice, maize, and cassava were cultivated and often mixed with *tumpangsari* (intercropping). Without the application of fertiliser the yield was low, and with no or only a short fallow period, the soil gradually became exhausted and taken over by *alang alang* grass. Only cassava could grow in this exhausted land, but the planting of this crop further reduced soil fertility. The poor families increasingly struggled to earn a subsistence level income. Unable to buy rice, these families turned to cassava as their staple food. Tapioca factories bought cassava at prices that were often so low that they only covered the harvesting cost. Working as wage labourers for estate plantations only provided a very modest supplement to their incomes. Estate plantations employed labourers only seasonally, such as during planting, weeding and harvesting, and because the wages paid were low, poverty has been a main characteristic of life among plantation workers.

Contract farming, a scheme involving smallholders ('the plasma') and estate companies ('the nucleus') was introduced in Lampung in the second half of the 1970s. Like the transmigration program, the main sponsor for this scheme was the World Bank. The aim of the contract farming program was to boost production and improve the livelihoods of the smallholder farmers. While production grew, the quality of life often did not. In Lampung, priority for the contract farming program was given to transmigrants. In Way Abung, for example, under the Perkebunan Inti Rakyat (People's Nucleus Estate) program, transmigrants were given credit and assistance by PT Perkebunan (PTP, the government-owned plantation company) to plant high-yielding rubber trees (*Hevea brasiliensis*) on their allotments. The plan was that transmigrants would secure their food supplies from dry or irrigated rice fields, while the rubber trees would provide additional cash. But eventually, instead of having two fields with one field for food supply and the other field to provide cash, the transmigrants either abandoned the rubber trees for food crops or abandoned the food crops for rubber trees (Levang 1989). Another example of contract farming was that of the transmigrant sites near the PTP Bunga Mayang sugar cane plantation and factory. The transmigrants were farmers who had been evicted from mountain forestry zones and forced to join the local transmigration program. Under the people's sugar cane intensification program (Tebu Rakyat Intensifikasi), the PTP gave credit and bought the sugar cane that the transmigrants planted on their land. The transmigrants' land titles were kept by the PTP as collateral. High debt, low yield and the low price of cane ensured a consistently low income among the transmigrants (Elmhirst 1997). The transmigrant families also supplied cheap seasonal labour for the cane plantation and sugar factory. Because the government only granted land title to the transmigrants, the neighbouring indigenous population was excluded from the project.

The latest schemes of contract farming in Lampung involved the production of cattle and shrimp. Central Lampung was the first to experience contract farming in cattle husbandry. Within the lucrative shrimp industries, two companies (PT Dipasena and PT Bratasena) were granted thousands of hectares of swamp and mangrove land between the mouths of the Way Mesuji, the Way Tulang Bawang, and the Way Seputih rivers. Under the people's nucleus shrimp pond program (Tambak Inti Rakyat), the companies recruited smallholders, provided credit, and bought, processed, and marketed the shrimp. The farmers provided the labour for the ponds' production. Each farmer's debt was deducted from the production under the promise that over a number of years they would be able to repay their debt and become the owners of the ponds. For example, in a contract between 8,600 families and PT Dipasena, they were promised that after eight years the farmers would be the owners of the ponds. But in 2000, after 10 years, not only did the farmers not own the ponds, the debt they owed was still extremely high. It was also estimated that the debt would take much longer to repay than had been originally negotiated (*Gatra*, 26 February 2000).

Converting swamps and mangroves into shrimp ponds is a more recent phenomenon. The surveys of Levang (1989) and Pain (1989) in the mid-1980s do not mention ponds as an important formal land use in Lampung. In the 1990s, most swamps and mangroves on the eastern and southern coasts of Lampung were gradually transformed into ponds. In addition to the two large-scale estate shrimp ponds, both under a contract farming scheme, medium-scale ponds can be found side by side with numerous small-scale ones. Like gardens, ponds provide wages for labourers. Ponds yield a lucrative profit, but unlike gardens and dry fields, the installation cost and input for the operation of an intensive pond is high. Thus the landless poor are unlikely to be able to afford to convert swamp and mangrove into ponds.

During the twentieth century, the opening of the 'empty land' of Lampung was completed. It was started in the early 1900s by the Dutch through colonisation, followed by the post-colonial transmigration programs, and was completed through the local transmigration programs.[11] The planned settlements started in the southern central region and continued northward encompassing the northeast Lampung lowlands. Spontaneous settlements that followed the same direction as the planned ones moved further toward the northwest Lampung highlands. The conversion of swamps and mangroves on the east coast into irrigated rice fields and shrimp ponds in the 1990s marked the last stage of the opening of the Lampung lowland. Forests in the northern part of the province were logged and converted into local transmigration settlements and large estate plantations. During the same period, the isolated mountain regions in

---

11   Between 1986 and 1988, Lampung sent 162 families to join transmigration programs in South Sumatra and Riau (Pain 1989: 317). The province never became a major transmigration area. By the 1990s, transmigration was no longer an important part of Indonesia's national development program.

the northwest (Sumber Jaya, Kenali, and Balik Bukit), named by Pain (1989) as Lampung's 'last frontier', continued to be cleared until the mid-1980s and were transformed into new population centres.[12]

By the very end of the twentieth century, Lampung had been transformed into an important producer of agricultural products. Pepper and coffee, of which Lampung remains Indonesia's centre of production, are still important commodities produced by smallholder farmers. Coconuts and bananas are largely produced from smallholder fields, though estate plantations also grow these crops. Despite being noted for centuries for its insufficient rice production and regular importation of rice from Java, Lampung has become a self-sufficient rice producer and is regarded as one of Indonesia's rice baskets. Among the annual food crops (for example, maize, soybean, peanut, and mung bean), cassava has become an export cash crop for which Lampung is the main national production area. Like sugarcane, cassava is produced by both smallholder farmers and estate plantations. Lampung has a surplus of livestock and the surplus is exported mainly to Java. Cattle are produced by small farmers, feeding companies, or under contract farming. Chickens are produced by small farmers as well as medium and large enterprises. Goats are raised primarily by small farmers.

In the wake of cash crop production by smallholder farmers and estate plantations (see Table 2-2), agricultural processing industries developed in Lampung. Besides dried and processed coffee and pepper, Lampung has become home to factories that process cane into molasses and sugar, and cassava into pellets and tapioca flour. Crude palm oil and crumb rubber have been exported from Lampung since the colonial period. More recently, processing factories in Lampung have begun to produce soap and detergent, monosodium glutamate, citric acid and sodium cyclamate.

Coffee and pepper processing factories are located close to the seaport at Panjang. Among these coffee processing industries, the Nescafé factory in Srengsem is the largest one. Feed processing factories are also located close to Panjang seaport. On the western and northern outskirts of Bandar Lampung, close to the old transmigration sites, are the oil palm factory in Rejosari near Natar and rubber factories in Way Lima, Way Galih, Bergen, and Bekri that have operated since the colonial era. Eight sugar factories are located between Gunung Sugih, Kota Bumi and Menggala. The cassava processing industry is the most dominant in the province with over 30 factories operating in 2000. Initially these factories were located near Panjang but have become scattered close to the cassava producing areas in East, Central, and North Lampung, as well as Tulang Bawang and Way Kanan districts.

---

12   Sumber Jaya, one of these new population centres, will be discussed at length in the following chapters.

Table 2-2: Land use and production data for Lampung Province, 2000.

Land use and commodities	Area		Production (tonnes)	Producers
	Hectares	(%)		
Rice fields	284,664	(8.6)		
Rice			1,992,689	Smallholders
Dry fields	675,860	(20.5)		
Cassava			3,613,919	Smallholders and plantations
Maize			1,109,326	Smallholders
Sweet potato			41,360	Smallholders
Soybean			12,024	Smallholders
Peanut			13,081	Smallholders
Green bean			6,352	Smallholders
Cash crop fields	1,031,811	(31.2)		
Coffee			95,165	Smallholders
Pepper			23,885	Smallholders
Coconut			139,617	Mostly smallholders
Cocoa			7,714	Mostly smallholders
Sugarcane			463,947	Mostly private plantations and smallholders[a]
Rubber			29,252	Mostly smallholders[b]
Oil palm			99,910	PTP and private plantations[c]
Forest	871,979	(26.4)		
Brackish pond (tambak)	33,844	(1.0)		
Alang-alang grass	90,164	(2.7)		
Settlements	248,109	(7.5)		

Notes: (a) 400,686 tonnes produced by private plantations, 5,500 tonnes by PTP, 57,761 tonnes by smallholders; (b) 22,988 tonnes produced by smallholders, 6,264 tonnes by private plantations; (c) 59,670 tonnes produced by private plantatuions, 40,240 tonnes by PTP.

Source: Biro Pusat Statistik (BPS) Lampung 2001.

By the end of the 1980s, the population of Lampung had reached the density of Java at the beginning of the twentieth century. With over 200 inhabitants per square kilometre in Java at that time, overpopulation was seen as the primary cause of economic stagnation and rural impoverishment. The distribution of its population to Indonesia's outer islands through the transmigration program was believed to be the solution. Although this belief has proved to be a fallacy (Wertheim 1959), the transmigration program has played a vital role in population distribution in Lampung (Benoit 1989; Pain 1989) and in boosting agricultural production. Population pressure however is not the only impact that the transmigration program (and regional development) has brought to

Lampung. Agrarian problems of rural impoverishment (such as low production per capita, landlessness, and poverty) are widely perceived to be linked to the overpopulation that characterised Java for centuries. This rural impoverishment has therefore also been successfully transmigrated to Lampung.

In the 1990s, Lampung was consistently ranked amongst the poorest provinces in Indonesia. In 1999 one out of every two families in Lampung was classified as poor (see Table 2-3). Poor families were classified as the ones who could not afford to live in proper housing and did not eat and dress properly. Particularly severe cases were those with a very slim chance of obtaining upward mobility. Lampung has indeed turned from the 'land of hope' into the 'land of despair'.

Environmental degradation also emerged as a problem facing Lampung. Waste from agricultural processing factories polluted the tributaries of the Way Sekampung, the Way Seputih, and the Way Tulang Bawang. Keeping the quality of the river water at a grade that it can still be used for agricultural purposes (irrigation and fishery) is one of the local government's main priorities. The conversion of swamps and mangroves into brackish shrimp ponds along the east coast is reported to have caused the erosion of beaches and the intrusion of saline water inland. Clearing of mountain forests by spontaneous migrant smallholders has for decades been understood to be a primary cause of watershed degradation. Hence forest squatters farming the upper watersheds that feed water into the big dams were the main target of government sponsored eviction and local transmigration programs. More recently, the remaining migrant smallholders farming the mountain zones have also been blamed for reducing the habitat of the endangered Sumatran animals such as the tiger, rhino, and elephant.[13]

Table 2-3: Population density and poor families in Lampung Province.

District	Area (km²)	Persons per km² (2001)	All families (1999)	Per cent poor families (1999)
Metro	78	1,501	26,165	16
Bandar Lampung	192	3,931	137,527	41
South Lampung	3,405	337	246,026	48
Tanggamus	3,401	235	181,335	45
Central Lampung	3,799	278	245,605	35
East Lampung	4,437	197	201,441	38
North Lampung	1,766	300	116,349	52
Tulang Bawang	7,770	92	165,004	68
Way Kanan	3,520	101	80,505	70
West Lampung	4,749	78	78,360	31
TOTAL	33,122	204	1,478,317	46

Source: BPS 1996, 2001.

---

13   Scientists from The Wildlife Conservation Society, a New York-based conservation organisation, suggest that the main threat that could lead to the extinction of these animals is the expansion of smallholder coffee gardens inside Bukit Barisan Selatan National Park boundaries (*BBC News*, 27 April 2004).

Conflict over land is another pressing issue in Lampung, where the incidence of land conflicts was among the highest in Indonesia (*Kompas*, 25 June 2001). There are conflicts between the local population and private and state-owned plantations and between the local population and the forestry authorities. A village in North Lampung, where local transmigration settlements were created near some longstanding Lampung hamlets, is a good example. Among the indigenous families in these hamlets there were disputes over which families had the right to receive the compensation given by a private company opening a plantation on their former traditional land (Elmhirst 1997). Some fields allocated to the local transmigrants were resumed by the indigenous Lampung people because they claimed they had not received compensation from the government. The local transmigrants found themselves among forest squatters who have been in conflict with forestry authorities for years. More recently, the indigenous communities have asked for compensation for thousands of hectares of their former traditional land now used by PTP Bunga Mayang for sugarcane plantations. Conflicts over forestry zones have been documented throughout Lampung. Suppressed during the New Order, the landless and near landless peasants took matters in their own hands after the *reformasi* and reclaimed the lands that were designated as forestry zones or granted to plantation companies.

The title of a national newspaper article, 'Land Conflict, Epidemic of Land Hunger' (*Kompas*, 25 June 2001), suggests that land hunger is one of the root causes of conflicts over land in Lampung. With a high agrarian density, a population of 400 persons per square kilometre, and 20–22 per cent of the population landless, land shortages have become a major problem in Lampung. As in Java, agrarian problems in this 'Little Java' or 'North Java' are perceived as the result of overpopulation.

As Table 2-3 showed, population distribution and regional development remain unevenly distributed. The city of Bandar Lampung, which is the capital's province, and Metro, a rural commercial centre newly classified as a municipality (*kota administratif*) are the most densely populated districts. This is followed by the early transmigration receiving districts (Tanggamus, South Lampung, Central Lampung, and East Lampung) and by the last local transmigration sites (Tulang Bawang and Way Kanan). With most of the region's mountain areas classified as state forest, significant areas are not available for settlement and therefore not designated as major transmigration receiving areas. West Lampung is the least populated district and has remained isolated until quite recently. Compared to other districts in Lampung, except for the Metro municipality, the least developed and least populated West Lampung has the lowest incidence of poverty (see Table 2-3). The chapters that follow will discuss West Lampung and one of its highland regions whose inhabitants have settled there quite recently. The highland region is regarded as 'the most developed area' in this underdeveloped district.

3. Creating a 'Wealthy Zone': Sumber Jaya and the Way Tenong Highland

Colonial and post-colonial government initiatives in the twentieth century brought mixed results in the Lampung Province in the form of poor zones in some areas and 'wealthy zones' in others. West Lampung was one of the province's least developed districts. However, a few regions in this 'undeveloped' district — Krui on the coast and Liwa (with adjoining Way Tenong and Sumber Jaya) in the eastern highlands — amply qualify as 'wealthy zones'. This chapter focuses on the creation of Way Tenong and Sumber Jaya which have become the province's most wealthy zones.

Indigenous populations are still relatively dominant in a number of coastal and highland regions of West Lampung. Non-Lampung migrant populations are highest on the southern part of the coast and in the eastern highlands. In ancient times, the West Lampung highlands were exclusively home to indigenous Lampung, but since the fourteenth century they have progressively left these highlands to settle the plains and coasts. A number of scholars argue that this out-migration is a result of the integration of the indigenous economy into world mercantilism.

In the eastern highland areas of Sumber Jaya and Way Tenong, the majority of the population is comprised of non-indigenous migrants. Migrants are primarily from neighbouring provinces: Semendonese from South Sumatra; Sundanese from West Java; and Javanese from Central and Eastern Java. In this chapter I argue that the in-migration of non-indigenous Lampung to this highland region can be linked to 'development' and the reproduction of smallholder farming.

I start by giving a brief history of the out-migration of indigenous Lampung from the West Lampung highlands in pre-colonial and colonial times. This is followed by an account of colonial and post-colonial in-migration of non-indigenous Lampung to Sumber Jaya and Way Tenong in the eastern highlands. I conclude by linking a description of the recent socio-economic conditions in Sumber Jaya and Way Tenong to the 'development' trajectory I have unveiled.

An Ancient Abandoned Highland: The Mountains of West Lampung

Lying between the borders of Lampung, Bengkulu, and South Sumatra in the north and the Sunda strait in the south, the West Lampung District can be divided into three geographic zones: Pesisir Krui forming the coastal strip; the southern hinterland and slopes facing the Indian Ocean to the west; and the mountainous highlands to the east. These gently rolling mountains and hills form part of the southern tip of Sumatra's Bukit Barisan mountain range which stretches the length of the island, from Aceh to Lampung.

Pesisir Krui is endowed with coconut groves and wet rice fields that dominate the narrow plains in the central portion of the coast. The southern coast also has upland fields comprised of annual crops (such as rice and maize) and, more recently, palm oil plantations. Cattle rearing is common in the region and damar tree agro-forests are present from the north to the south of Pesisir Krui, dominating the slopes up to an altitude of 800 metres. Here, along with other fruit and timber tree crops, indigenous smallholders cultivate *Shorea javanica* trees following successions of rice swidden with coffee and/or pepper gardens (Michon et al. 2000).

In the highlands, Mount Pesagi reaches 2,239 metres above sea level. Most of the surrounding mountains and hills are classified as forest reserves. Patches of forest can still be found on the upper slopes or on the tops of mountains and hills. Some villages have protected patches of forest adjacent to wet rice fields and settlements. Most settlements are located between 700 and 1,000 metres in elevation. Smallholders cultivate coffee, pepper, and other tree crops in the highlands. Terraced wet rice fields are constructed on the alluvial flats adjacent to creeks and rivers.

The highlands of West Lampung have become home to both indigenous communities and migrant populations of Semendo, Javanese, and Sundanese. The indigenous Pesisir population is dominant in the western part of the highlands, including the regions of Sukau, Balik Bukit, Belalau, and Kenali. In the eastern part of the highlands, numerous old Semendo villages can be encountered in Way Tenong, but not so many in Sumber Jaya. Sundanese and Javanese hamlets and villages can be found almost everywhere in the West Lampung mountains. The concentration of hamlets and villages of migrants from Java is increasing as mountain areas such as Sekincau and Suoh in the east are newly cleared. Migrant populations (Semendo, Javanese, and Sundanese) represent the majority in this 'newly developed' region of Sumber Jaya and Way Tenong in the eastern-most regions.

The early history of the West Lampung Highlands identifies a flourishing ancient civilisation. Scattered megalithic remains can be found in the highlands. Batu Brak, the largest site of these megalithic remains, is located in Kebon Tebu, Sumber Jaya. In the centre of an area of about 2 hectares, a menhir or large standing stone is circled by neatly laid dolmens. In addition to megalithic stones, a series of archaeological excavations have also found bronze bracelets, blades, beads, and shards of locally made and imported pots. Sukendar (1979) interprets the artifacts as representing ritual objects used in burials and religious worship as well as for more mundane uses such as food processing, tool making, and building materials. According to McKinnon (1993), the shards of ceramics, thought to have been imported from China during the ninth and tenth centuries AD, indicate that foreign trade was occurring in these highlands in ancient times.

The relationship between the ancient communities of Batu Brak and its neighbouring megalithic sites and the present people of Lampung is not well established. One thing that is reasonably certain, however, is that the disappearance of this ancient civilisation permitted the modern day population to migrate and settle in the West Lampung Highlands.

A more recent in- and out-migration history of Lampung suggests that the West Lampung highland region was abandoned by its population (Hadikusuma 1989; Sevin 1989). The majority of the present-day indigenous groups trace their origins from the West Lampung Highlands. Sekala Brak, a location in the foothills of Pesagi Mountain near Lake Ranau, is said to be their land of origin. Different periods and directions of migration have resulted in different dispersal patterns of indigenous Lampung populations (see Sevin 1989). Based on oral and written histories of indigenous communities collected by Dutch scholars and officials, it is thought that the first waves of out-migration took place during the fourteenth and fifteenth centuries. Groups from the highlands moved to the central and eastern plains where they developed as a sub-group of indigenous people known as the Abung. A second and subsequent wave of migration dispersed to the southern and western lowlands and coast. In the eighteenth century they were identified as Pesisir (or Peminggir). The out-migration of the Pesisir from Belalau continued up until the twentieth century. Both Abung and Pesisir later either absorbed or drove out the Pubian, the third and smallest group of indigenous people living in the central and southern Lampung plains. Unlike Abung and Pesisir, Pubian oral history does not strongly link their origins to the Belalau highlands.

Subsequent waves of migration from highland to lowland Lampung are thought to be linked to pre-colonial and colonial mercantilism and the characteristics of indigenous social organisation.[1]

---

1   A history of Lampung before the migration from highlands to lowlands is difficult to ascertain. Historical materials provide convincing evidence of the existence of an earlier civilisation in lowland Lampung (Hadikusuma 1989). A Chinese source indicates trading relations between China and Tulang Bawang on the north coast as early as the seventh century. Stone plaques describing the Sriwijaya's power and influence in Lampung at the end of the first millennium were found in several places. Signs of the presence of Majapahit in Lampung in the thirteenth century can also be traced.

Between the sixteenth and eighteenth centuries, the Sultanate of Banten — then the world's primary pepper supplier — obtained pepper supplies from Lampung. From the eighteenth century, the Dutch obtained pepper supplies directly from the eastern portion of Lampung. The British controlled the pepper supply from Lampung's west coast from the late seventeenth to the early nineteenth century. This included the present-day West Lampung District, which then included part of the residency of Bengkulu. During the second half of the second millennium, the indigenous Lampung population was the most important global pepper producer.

A British report written in 1813 (Bastin 1965: 147–8) notes that on the west coast of Krui, 881 married men and 640 single men were engaged in an informal 'contract' with the British to farm various stages of pepper gardens. These men maintained almost half a million pepper-bearing vines and an equal number of non-bearing vines (newly planted and old). The production for that year was 147.6 tonnes. In addition, there were another 119,550 bearing vines that produced 24 tonnes of pepper in 'free' gardens. An earlier historical record — a seventeenth century plaque — indicates a similar contract between the indigenous Lampung producers on the southern coast and the Sultan of Banten (Kingston 1987: 10–1). A married man was expected to plant 1,000 pepper vines while bachelors were to plant 500. By buying the pepper at a set price, the Sultan monopolised sales and claimed a minimum of 11 per cent as tribute.

Up to the mid-nineteenth century, the Sultanate of Banten, the Sultanate of Palembang, and Bugis and Malay traders were involved in a series of conflicts with pirates downstream of Way Tulang Bawang in the northeastern part of Lampung. Control over pepper produced in the surrounding areas was at the heart of the conflict. From evidence of pepper trading in the lowlands of Lampung, it can be assumed that pepper cultivation may well have been a motivation for the migration of indigenous peoples from the highlands to the lowlands.

If engaging in petty commodity production for global trade inspired the indigenous Lampung to move to the lowlands, the process was also mediated and even facilitated by customary practices such as marriage, property, inheritance and other socio-political structures. Payment of a high bride price was a prominent characteristic of marriage among the Lampung people (Wilken 1921, cited in LeBar 1976). The indigenous Lampung practised virilocal post-marital residence and male primogeniture in inheritance. The bride was 'taken' from her group and the children 'belonged' to the groom's group. House and land passed to the elder son who was then responsible for the care of the parents and unmarried siblings. The size of the brideprice and the marriage party was negotiated in accordance with the status of the family in the community. The higher the status, the higher the brideprice payment. Larger wedding parties required more buffaloes to be slaughtered and more meals to be served. Having inherited none of their parents' property, after marriage the younger brothers worked on their own farms to provide their families with food, a sturdy house, and enough resources to pay for the brideprice and the wedding party when their sons got married. Pepper cultivation and, later in the nineteenth century, coffee production enabled this system to persist. New land was constantly sought for pepper gardens. Forests were cleared for upland rice swiddens in the first year or two and transformed into pepper gardens (and/or coffee gardens later in the nineteenth century) to be managed for another ten years or more. Old gardens that had been left fallow were later rejuvenated, transformed into tree gardens, or simply abandoned for natural regeneration. A new forest plot was cleared and the cycle of such rotational cultivation continued.

A dominant tradition among Lampung communities occupying new territory involved a process of political fission. As discussed in Chapter Two, *buay* and *marga* are recognised as the largest socio-political units of the indigenous people. Each *marga* was independent of other *marga*.[2] Rather than uniting into a single kingdom, it is evident that the indigenous Lampung were continuously creating independent *marga*. This typically took place when groups of people migrated to establish new gardens and create new villages on land beyond the boundary of their mother *marga* territory. With established trading networks for pepper on the coasts (Krui and Semangka Bay) and the presence of navigable rivers such as the Way Tulang Bawang in the north, the Way Seputih in the centre and the Way Sekampung in the south, lowland Lampung attracted more and more migrants from the highlands.

The waves of migration of indigenous people from the highlands to the lowlands eventually left extensive tracts of the West Lampung highlands 'unpopulated'. In the early nineteenth century, a few small villages surrounded by mountain forests were scattered in the regions of Balik Bukit, Belalau, and Kenali. As noted in Chapter Two, by the mid-1800s, the Dutch had gazetted the non-cultivated lands between settlements and fields as state property. On one hand, this action limited indigenous people's access to forest land between their settlements and fields, but on the other hand it enabled the Dutch officers to allow migrants to move in and occupy former indigenous *marga* lands.

The present day mountain region of Way Tenong and Sumber Jaya — then known as the territory of *marga* Kenali — became an 'empty' frontier. It is this empty land that attracted an influx of more recent migrants, this time from outside Lampung.

---

2   Some of the independent adjacent *marga* formed loose confederations, such as Megou Pak (the four *marga*) on the southern coast (that later supported Raden Intan, his son Raden Imba Kusuma, and his grandson Raden Intan II's rebellion against the Dutch in the 1800s), and Abung Siwo Mego (the nine *marga* of Abung) who all claim to be descendents of the same mythical ancestor Minak Paduka Begaduh, a migrant from Belalau.

Map 3-1: Sumber Jaya and Way Tenong.

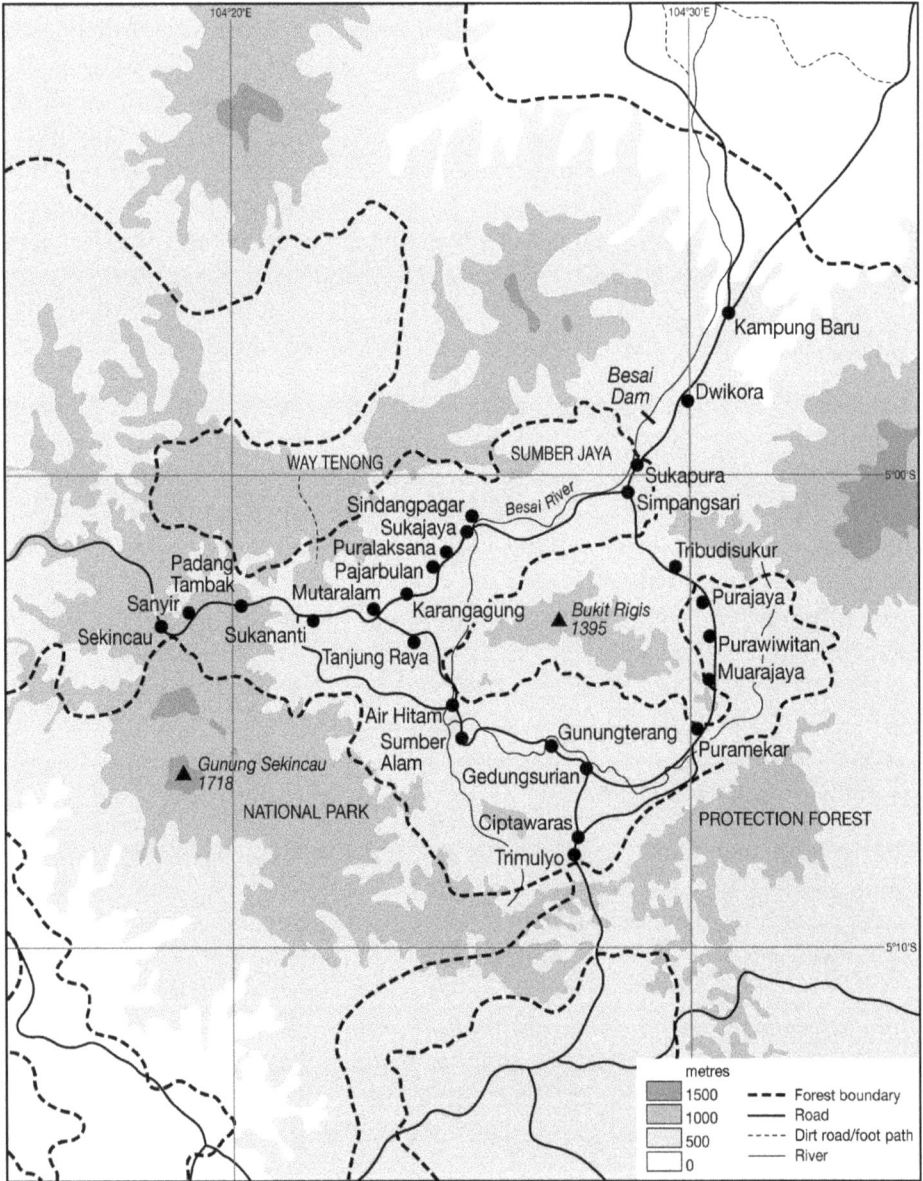

Source: CartoGIS, ANU.

The Coming of the Semendo: Way Tenong

Semendo is the name of a sub-group of Pasemah people inhabiting highland Palembang in the province of South Sumatra.[3] Compared to other sub-groups of Pasemah, the Semendo were said to have their own distinct characteristics of social organisation (LeBar 1976). While other Pasemah sub-groups are organised genealogically into patrilineal clans (*sumbai* or *marga*) and lineages (*jurai*), the Semendo have matrilineal clans and lineages. Other Pasemah sub-groups practised the prevalent system of marriage involving a high bride price, virilocal post-marital residence, and male primogeniture for inheritance. In contrast, Semendo marriage involved no brideprice payment, uxorilocal post-marital residence, and female primogeniture for inheritance (*tunggu tubang*). The *tunggu tubang* stipulates that the eldest daughter inherits the parents' property, usually the house and land. The Semendo, among the Pasemah, were also the earliest to convert to Islam, and their wet rice fields were more advanced than those found anywhere else in southern Sumatra in the nineteenth century.

An impetus for the migration of the Semendo can be attributed to the practice of *tunggu tubang*, which forced residents to look for new land to clear elsewhere (Sevin 1989: 93). Within the Pasemah land, the Semendo first migrated to Semendo Ulu Luas and Mekakau, and later moved further down to Bengkulu and Lampung. In the 1870s the Semendo started their subsequent southward migration to Lampung. The Semendo first moved to present day Kasui, Way Tenong, Sumber Jaya, and Pulau Panggung, migrating along the eastern slopes of the Bukit Barisan mountain range. They cleared the forest, created villages and wet rice fields, and opened upland rice fields that were then transformed into coffee gardens that were often inter-planted with pepper. The Semendo established four 'independent' *marga* in the 1930s along this route of migration. These *marga* (from north to south) are Kasui, Rebang Seputih, Way Tenong, and Rebang Pugung.

It is important to note that colonial interventions facilitated the further southward migration of the Semendo people into Lampung. By the 1850s, the Dutch had been able to place the territory and the people of Palembang, Bengkulu and Lampung under their political control. All of the villages and *marga* in these three residencies were integrated into the colonial government administration. Using these three villages, the Dutch overthrew the British-controlled Singapore trading networks, and reoriented the trading of commodities (especially pepper and coffee) via Batavia (Jakarta) as an obligatory transit.[4] Migrating to

---

3 According to Jaspan (1976), the Pasemah in 'a broad sense' include the 'linguistically kindred' groups of Empat Lawang (Lintang), Gumai, Kikim, Kisam, Lembak, Lematang, Mekakau, Pasemah Lebar, Semendo and Serawai. In 'a strict sense', the term Pasemah refers only to the people of Pasemah Lebar.

4 In the middle of the nineteenth century, pepper was no longer the only commodity sought from Lampung and production decreased to only 10 per cent compared with a century before. The Dutch liquidated the VOC

Lampung to get closer to the trading posts in Semangka Bay therefore offered an economic advantage to the Semendo. In the 1850s, the Dutch imposed a new system of land ownership (Kingston 1987) that enabled the Semendo people to occupy land in Lampung. The government only recognised land claims by individual villages up to six kilometres from the village and three kilometres from a temporary hamlet on newly cleared land. The land located between the villages, formerly common *marga* territory, now became a state domain. The Dutch administration allowed non-Lampung migrants to occupy and settle on some of this newly gazetted 'public land', which led to the Lampung *marga* no longer being in a position to protect the traditional claims of its members to frontier land (ibid.: 242) or to resist migrants seeking to settle and farm their former common land.

During the Dutch administration, West Lampung District was known as the 'sub-division' of Krui and formed part of Bengkulu residency (see Sevin 1989). Of the four Semendo *marga* in present day Lampung, Way Tenong formed part of the sub-division of Krui under the Bengkulu residency administration. Elders in Way Tenong often reflect on the story of the first migration of Semendo to Way Tenong. One version of this story, as told by Pak Jahri, the former village head of Mutar Alam, was published as a 'brief history of ex-*marga*, Way Tenong':

> In 1884, a group of men, Imam Paliare (Abidun), Raje Kuase (Serimat) and Puting Merge (Sendersang) and their followers Jenderang (Buntak), Jemakim, Senikar and Jakalam received an order from Puyang Awak to search for land around the headwaters of Way Besai River. These men lived in *marga* Ulu Nasal in Bengkulu. They were told that Way Besai was located in Rantau Temiang. So they went to the village of Rantau Temiang in Rebang Kasui. When they arrived there, two persons, Panjilam and Sersin, welcomed them. They continued travelling along the Way Besai River and stopped at Gedung Aji, now the site of the Way Besai hydroelectric power plant. In 1885, at Gedung Aji, they cleared the forest and opened upland rice fields for a year while continuing the search for the head of the Way Besai River. After a year, in 1886, they finally found the location they were looking for and moved there. They called this newly cleared land Mutar Alam.
>
> After building a settlement in Mutar Alam, they travelled back to Rantau Temiang in Rebang Kasui and continued to Menggala to seek permission [to create the village administration] from the Dutch officer. In Menggala they reported to the officer their new location at the head

---

at the end of the eighteenth century. Coffee, among other cash crops (such as sugar and pepper), was planted by peasants under the system of 'forced cultivation' and by private companies in parts of Java, Sulawesi and Sumatra. By the twentieth century, in the southern half of Sumatra, coffee in the highlands and rubber in the lowlands became an important source of income for smallholder farmers.

of the Way Besai and asked for permission. They were told that the land at the head of Way Besai was not under Menggala administration, [and that] the land was under the jurisdiction of the Department of Krui. The delegates were given an official letter to report to *marga* Kenali. In Kenali, the delegates met the chief of the *marga* Pangeran Polon. He accepted the new settlers as residents of *marga* Kenali. He appointed Puting Merge as the head of the new settlement, [who was] to report to him every three months about the development of the population and to receive further instructions.

As the population grew, new hamlets were created. In 1887, the new hamlets included: 'old' Fajar Bulan (now Sukajaya), Karang Tanjul (now Karang Agung), Gedung Surian and 'old' Sukaraja. In 1891, the resident of Bengkulu officially recognised all these hamlets as parts of the administrative village of Mutar Alam and appointed Serimat as village head.

In 1900, after a long approach to *buay* Belunguh and *marga* Kenali, the status of *marga* was finally granted. To mark the separation of Way Tenong (the name of the new *marga)* from the *marga* of Kenali, a set of gifts was given by the new community to the *marga* Kenali. The gifts included a sum of cash, a buffalo, a hundred dishes of rice cooked in sweet coconut milk, a hundred dried/fermented semah fish and an elephant tusk. The two *marga* were declared as siblings (*kakak adik*), with Kenali as the elder and Way Tenong the younger. The boundary of the territory of the new *marga* was then set. The boundaries were Air Sanyir/Sekincau to the west, Dwikora to the east, Mount Remas to the north and Begelung Ridge to the east. Also [that year], the Krui Dutch officer officially appointed Raden Cili as the first *marga* chief (*BUMIpos*, 11 September 2000).

According to many elders, the common pattern of creating new settlements was for small groups of families to depart from their village and clear new forest areas for cultivation. They sought fertile and relatively flat land where water could be channelled for wet rice fields. When this land was found, the forest was then transformed into permanent agricultural fields. This endeavour by a group of families to find new land to farm was called *nyusuk*. The cleared land evolved from a hamlet or village with a few scattered houses and huts, to settlements usually organised along the main road or path. The first land cleared in Way Tenong was the old hamlet and wet rice fields in the village of Sukaraja spread out over approximately 40 hectares. The fields were cleared and distributed among the first group of families arriving from Ulu Nasal, Bengkulu. Villages in the area were comprised of rows of old stilted wooden houses along the main road near the wet rice fields, following the banks of Way Besai River and its tributaries. It is said that the cultivation of coffee was initiated later after the Dutch agricultural officers informed the people that the soil was suitable for coffee and advised them to plant this lucrative export crop. Coffee was then planted — with and without the initial one or two crops of upland rice — in the upland after the forest had been cleared. After 15 to 20 years, the fields were left fallow.

In the first half of the twentieth century there were five Semendo villages in Way Tenong: Sukaraja; Mutar Alam; Gunung Terang; Karang Agung; and Way Petai (Pain 1989: 304). In the 1950s, when transmigrants from the National Reconstruction Bureau (Biro Rekonstruksi Nasional) created new villages and a separate administrative sub-district, all of the villages in *marga* Way Tenong were integrated into the new sub-district of Sumber Jaya. Simpang Sari, the capital of the new sub-district of Sumber Jaya, is much closer than Liwa, the capital of the sub-district of Balik Bukit to which Way Tenong formerly belonged. It took a day's motorbike travel to go to Liwa, but only an hour or two to travel to Simpang Sari.

When discussing their traditions, the Semendo in Way Tenong and Sumber Jaya will mostly refer to *tunggu tubang* where the parental house and land is inherited by the eldest daughter who, in return, is responsible for the care of her parents. Those who have no daughter bequeathed their property to their eldest son. This less preferred practice is called *nangkit*. Selling the *tunggu tubang* house and land is unacceptable and very rare. Thus, one can easily find in the region many *tunggu tubang* houses, wet rice fields, and coffee gardens, some of which have remained intact for four generations while new ones are continually created. Old men usually relate the concept of *tunggu tubang* to politeness between men and women (*singkuh sinduh*). To live with your own daughter in the same house is more acceptable than to live with your daughter-in-law. For example, it is extremely impolite for a man to be at home only with his daughter-in-law, to eat alone in the kitchen with his daughter-in-law, or even to be fed by his daughter-in-law when he is sick.

Semendo in the region also pay special tribute to their ancestors (*puyang*). Many people believe that the Semendo in the region are descendents of the mythical ancestor Puyang Awak, who is said to be 'immortal'. Puyang Awak is believed to be immortal because he has no grave and his whereabouts are unknown. Great-grandparent's graves are cared for and frequently visited for prayers (*ziarah*). In the villages of Mutar Alam and Gunung Terang, a ritual feast of *sedekah pusaka* (feast to celebrate ancestors) is held each year in the Islamic calendar month of Muharam. In these ritual gatherings, the descendents of the 'founders' of the villages — a male in Mutar Alam and a female in Gunung Terang — recite verses from the Qur'an and pray for their ancestors. In both villages the *sedekah* is also marked by the cleaning of a dagger heirloom (*pusaka*) and concluded with a meal attended by the entire village.

An Enclave of Indigenous Lampung: Muara Jaya Village

Muara Jaya is the only village in the Sumber Jaya and Way Tenong region that is almost exclusively populated by indigenous Lampung Pesisir. Surrounded by Semendo, Sundanese, and Javanese villages, Muara Jaya is now an enclave. In this village, there are no more than about 200 Pesisir families. Amongst themselves, the indigenous people in Muara Jaya still use the Pesisir dialect of the Lampung language even though some Javanese men and women have inter-married with them. Like all Semendo villages, the majority of Pesisir families in Muara Jaya live in stilted wooden houses.

The Lampung Pesisir population in Muara Jaya first arrived in 1930 when seven families moved from Sebarus in Liwa. The land was inside the territory of the Way Tenong *marga* so they needed permission from the Semendo people. The Semendo of Gunung Terang village were consulted and gave them permission to clear the land and settle in their present location. A year later, these seven families returned to Liwa immediately after a large earthquake and, in the years that followed, some of these families (together with new families) came to Muara Jaya. In 1949 the new hamlet of Muara Jaya was officially acknowledged as an administrative village. In the mid-1990s, a section of the village with relatively few indigenous Lampung was officially recognised as a separate village, so now there are Muara Jaya I and Muara Jaya II.

According to elders in Muara Jaya, looking for new land for wet rice fields was the primary reason for their migration from Liwa. The alluvial riverbank flats suitable for wet rice fields were a source of conflict in the 1950s and 1960s between the Lampung and the neighbouring transmigrants. Both groups claimed ownership over the same land, which was considered 'precious' by both groups. The dispute was resolved after high profile mediation by the provincial and national authorities.

Apart from wet rice fields, the Lampung also planted upland rice (*padi ladang* or *padi darat*). Some elders also said that they had heard that the Dutch administration planned to open a tea plantation in the region, but this plan never materialised. After the arrival of transmigrants from Java, it was said that coffee became a significant source of income during the 1950s. It is important to note that the world-wide economic depression in the 1930s, followed by Japan's occupation of Indonesia in the first half of the 1940s, and Indonesia's war of independence against the Dutch in the second half of the 1940s, caused the decline of markets and smallholders' production of cash crops, including coffee. Consequently, during the 1930s and 1940s, rice production from wet and dry/upland fields became the primary subsistence product for Muara Jaya villagers' and others in the archipelago. Indonesian Independence, declared in 1945 though not acknowledged by the Dutch until 1949, marked the revival of coffee production in the region. Transmigrants from Java and subsequent developments have further facilitated this revival.

The Arrival of Transmigrants from Java: The Creation of Sumber Jaya

Unlike transmigration projects elsewhere in Indonesia, which are organised by the Office of Transmigration, the transmigration project in Sumber Jaya and Way Tenong was organised by a special unit under the office of the then Prime Minister of Indonesia. This special unit, called the Biro Rekonstruksi Nasional (BRN), was designed to assist soldiers and civilian militia who had been involved in the war of independence. The assistance was considered a kind of reward for these freedom fighters and was primarily aimed at ensuring their return to 'a normal life'. One obvious choice was to turn these fighters into smallholding farmers by allotting each of them a piece of land. Since there was no more land to be distributed in Java, they had to be transmigrated outside of Java. Lampung was chosen as the destination due to its location close to Java and previous experience with receiving transmigrants. Several locations in Lampung were selected to receive the BRN transmigrants, and the 'empty' Way Tenong highland area was one of these.

Mimicking the structure of the army, the BRN transmigrants were organised into groups, each under the leadership of a commander (Hereen 1979). Under this leadership, each separate group cleared the forest, built a housing compound and road, and cultivated the land. Through their group leader, the transmigrants received government assistance in the form of cash, food, tools, and building materials in the initial years. From 1949 to 1959, seven new transmigrant villages were created. The first locations to be cleared were the present villages of Sukapura and Simpang Sari to the east of Bukit Rigis Mountain. From here, clearing continued to an area called Kebon Tebu to the south of the same mountain, where three villages were created (Tribudi Sukur, Pura Jaya, and Pura Wiwitan), and up to the northwest of the mountain, where two villages were created (Fajar Bulan and Pura Laksana), close to the Semendonese villages in Way Tenong.

Most BRN transmigrants were Sundanese and were from different parts of West Java such as Tasik Malaya, Karawang, and Bogor. There were few Javanese. It is interesting to note that the number of actual veterans was very small. The implication is that most BRN transmigrants to Sumber Jaya had likely never been involved in the independence war, and that more than half of the migrants were actually farmers and labourers (Heeren 1979: 72). There are no precise data on how many 'official' BRN transmigrants arrived in Sumber Jaya. The BRN office recorded 22,198 members transmigrating to Lampung during 1951–53, among them 9,205 persons (2,441 families) who transmigrated to North Lampung, while the rest went to other sites in south and central Lampung (ibid.). In North Lampung there were two BRN sites — Sumber Jaya and Tanjung Raya. The latter consisted only of one village in 1952, but a decade later had developed to include two other villages (Sevin 1989: 107). Heeren (1979: 81–3) noted that Sumber Jaya was the largest BRN transmigration site in Lampung. Transmigrants in Sumber Jaya were organised into two main organisations — Loba and Pencak Silat (or PS), the latter being further divided into PS51, PS52, and PS53. The Loba members settled in Sukapura.[5] The PS51 group occupied Simpangsari, and 450 families arrived in 1951, but by 1954 only 115 of them remained. By 1957, there were 715 families in the PS52 and PS53 groups, of whom 2,592 people (in 12 sub-groups) lived in Kebon Tebu, while 2,029 lived in Way Tenong.

Heeren (1979: 81–93) further notes the development of cooperatives among BRN transmigrants in Sumber Jaya, as well as problems with the neighbouring Semendo and Lampung people during the period from 1951 to 1957. Under the organisation of Loba and PS, the transmigrants developed cooperatives for production and consumption. The land was cleared, cultivated, and harvested collectively. All of the harvests 'belonged' to the organisation and each member received food, goods, and a small amount of cash for their daily needs. The harvests were sold and the surplus kept by the organisation, allowing it to ensure that all of its members had enough food to eat. Houses were built collectively. For the first five years, the land and houses could be individually owned but to sell them was prohibited. Hereen suggests that under Loba, the development of the cooperative was very positive, with the organisation owning six shops, a sawmill, and a tile factory. In contrast, the PS cooperatives in Kebon Tebu were soon in a state of crisis. Here harvests had failed and roads were not properly maintained. Collective farming soon gave way to individual production. With regard to the development of cooperatives, the success of Loba and the failure of PS has been largely attributed to the skills and qualities of the local leaders. Loba had strong, charismatic and reliable leaders, while the PS did not.

In Sukapura and Simpang Sari, the average size of land holdings was 1.1 hectares per family, while in Kebon Tebu it was 0.8 hectares. Both of these figures were far below the ideal and planned average of 3 hectares per family (Hereen 1979). Besides rice, the transmigrants cultivated maize, potato, cabbage, European vegetables (like cabbage and carrot), coffee, and a small amount of pepper. Since the road had not yet been constructed in those initial years, transporting these commodities was the main constraint.

---

5   An elder in Sukapura said that there were about 400–600 families in Loba, many of them from Tasik Malaya. Some of the Loba members later created the separate village of Tribudi Sukur.

Claims and counter-claims over land between the transmigrants and the neighbouring Semendo and Lampung people constituted another problem. There were cases where the indigenous Lampung and Semendo settlers claimed land that had been transformed into irrigated rice fields by the transmigrants. These conflicts were largely due to the fact that, unlike other transmigration sites elsewhere, in Sumber Jaya the transmigration project was not preceded by the process of field delineation to define the boundaries of the land allocated for the transmigrant villages.

Also in the 1950s, the BRN transmigration villages created a separate administrative sub-district (*kecamatan*) and refused to be integrated into the existing administrative sub-district (*negeri*) of Balik Bukit. The transmigrants' concern was that under the Balik Bukit *negeri* they would be an inferior minority 'ruled' by Lampung and/or Semendo administrators. By creating a separate *kecamatan*, the BRN transmigrants were able to interact directly with higher level authorities with a better chance of persuading them to bring village development projects to their newly created village of 'freedom fighters'.

Then Indonesian President Sukarno and Vice President Hatta officially inaugurated the formation of Sumber Jaya as an administrative *kecamatan* in 1952. Elders in Sumber Jaya hold the memories of Sukarno and Hatta's visit to Sumber Jaya dear. It is said that the president himself chose *sumber jaya* ('source of glory') as the name for the new *kecamatan*. Sukarno's speech transcript, a hand-written plaque, and a photograph are preserved commemorating the occasion. The president also laid the first stone foundation for a monument named in his honour (Tugu Sukarno) in Simpang Sari. A hamlet in Sukapura is named Sukarata after Sukarno and Hatta. The wooden house in Simpang Sari where both men stayed during the visit has been preserved.

The Flood of Spontaneous Migrants

The Semendo from the neighbouring Way Tenong and Kasui areas were quick to decide 'to get closer' to these transmigration villages, and literally did so by clearing the land adjacent to these new settlements. While aligning themselves with transmigration settlements as an initial strategy to benefit from government development projects, the Semendo had a more dramatic next strategy that involved bringing Javanese and Sundanese migrants to their villages. In this way the Semendo villagers hoped to receive government programs and projects similar to those of the transmigration villages. It was this pattern that later brought a flood of many more spontaneous migrants to the region. Through this strategy, the number of villages in the region doubled in three decades. Thirteen villages (five of Semendo transmigrants, one of Lampung, and seven of Sundanese and Javanese BRN transmigrants) in the early 1960s grew to

26 villages by the mid-1980s (Sevin 1989: 304). The Semendo and spontaneous migrants — most of whom were Javanese — later created ten new villages. Of these ten villages, four (Padang Tambak, Suka Menanti, Tanjung Raya, and Sindang Pagar) were populated by both Semendo and Javanese migrants, while the other five (Sidodadi, Sri Menanti, Sumber Alam, Tri Mulyo, and Gedung Surian) were populated mostly by Javanese migrants. In addition, the BRN transmigration villages created three more administrative villages — Pura Mekar, Cipta Waras, and Sukajaya.

It is interesting to note this new approach by the Semendo villagers. Not only were more and more Javanese and Sundanese migrants welcomed to settle in their villages, but part of their village land was allocated to the new migrants for the creation of new villages. Not all of these Javanese and Sundanese migrants came directly from the island of Java; many were born or had lived in old transmigration sites in south, central, and north Lampung. In many cases, these Javanese migrants were given the land for free. For example, the village of Gunung Terang gave part of its still forested village territory to groups of Sundanese and Javanese migrants. These groups then later created the villages of Gedung Surian, Cipta Waras, Trimulyo, and Semarang Jaya. In the same way, the village of Sukamenanti gave and sold land to Javanese migrants to create Sidodadi and Sri Menanti, where migrants transformed the forests and bush into villages. The Javanese migrants were also welcomed in established Semendo villages. They could work farming the unused plots (*numpang*), as sharecroppers (known locally as *garap*, *maro* or *bagi hasil*) or wage-labourers (known locally as *bujang* or *upahan*) until they were eventually able to buy a piece of land of their own. Usually the land was bought through a series of small payments (*cicilan*) at the end of the coffee harvest season.

The reason that the Semendo were so generous in giving land to Javanese migrants, apart from obtaining abundant labour for their coffee gardens and wet rice fields, was to attract government programs and projects such as roads, schools, clinics and markets. According to the former heads of these Semendo villages, the arrival of the Sundanese and Javanese brought progress to their people. As these former village heads put it, 'without the migrants from Java, there would be no development projects and no progress in our villages'. The result of this approach was that more and more migrants arrived, more administrative villages were created, and there was more 'development' and 'progress' in the villages and in the region. The region was soon transformed into a 'wealthy' flourishing highland region providing migrants with opportunities for a better life. Many did attain a better life, but others certainly did not.

A Flourishing Highland

During my research, when someone visited West Lampung District and asked where are the 'fastest developing regions' (*daerah yang paling cepat maju*), the answer most likely was Sumber Jaya and Way Tenong. In the easternmost part of the district, the two capitals of these regions — Sumber Jaya (the capital of Sumber Jaya sub-District) and Fajar Bulan (the capital of Way Tenong sub-District) — were flourishing. The region had become the new commercial and population centre in the Lampung highlands and one of Lampung's most important 'coffee pots'. The region was dominated by smallholder agricultural production. The final part of this chapter elaborates on the socio-economic conditions in this flourishing region.

A Developing Region in an Underdeveloped Country

The level of 'advancement' of Sumber Jaya and Way Tenong is particularly meaningful in the context of modern Indonesia. During the Suharto New Order period (1966–98), development (*pembangunan*) and progress (*kemajuan*) were key words in the grand project of Indonesian nation building. Analysing how much progress a particular region had made and which particular region within a certain administrative boundary was the 'fastest developing' was seen as a key measure of the success (or failure) of a region.

One indication of progress in the Sumber Jaya and Way Tenong regions was the absence of IDT[6] or 'left-behind' villages within its boundaries. A village was classified as an IDT village if it lacked the facilities and services (for example, roads, schools, health clinics, and markets) found in the average village in the province. In the mid-1990s, only two out of over 24 villages in Sumber Jaya and Way Tenong were classified as IDT villages. This was much less than the average for West Lampung District, which was one out of every two villages (see Table 3-1).

Table 3-1: Total villages and 'left-behind villages' in Lampung Province.

Sub-districts	Villages in 1996	IDT villages in 1996	
		No.	%
Metro and Bandar Lampung	84	3	3.6
South Lampung and Tanggamus	642	226	35.2
Central and East Lampung	516	91	17.6
North Lampung, Tulang Bawang, Way Kanan	579	284	49.1
West Lampung	162	80	49.4
TOTAL	1,983	684	34.5

Source: Badan Koordinasi Keluarga Berencana Nasional [National Coordinating Office for Family Planning] 2001.

---

6   The acronym derives from Instruksi Presiden Desa Tertinggal (Presidential Instruction on Left-Behind [or Neglected] Villages).

Thanks to a number of subsequent poverty alleviation and rural development projects, the number of IDT villages in West Lampung District had gradually been reduced to almost half, from 80 (49 per cent) out of 162 villages in 1996 to 42 (25 per cent) out of 169 villages in 2000. In 2000–2001, it was only in Balik Bukit where Liwa, the capital of West Lampung District is located, that there was a complete absence of IDT villages other than in Sumber Jaya and Way Tenong (see Table 3-2). The absence of IDT villages in these three sub-districts suggests, albeit in a narrow sense, that progress has been achieved.[7] It also reflects a lack of such progress in other sub-districts.

Table 3-2: Population, poor households, and 'left-behind villages' by sub-district in West Lampung District, 2000.

Sub-district	Households			Villages		
	Total	Poor	(%)	Total	IDT	(%)
Bengkunat	7,562	4,006	(53)	16	6	(38)
Pesisir Selatan	3,875	1,348	(35)	10	1	(10)
Pesisir Tengah	5,946	1,183	(20)	20	2	(10)
Karya Penggawa	2,611	384	(15)	8	2	(25)
Pesisir Utara	2,015	356	(18)	16	7	(44)
Lemong	2,896	612	(21)	11	7	(64)
Sukau	5,346	224	(4)	9	2	(22)
Balik Bukit	5,193	1,497	(29)	11	0	(0)
Belalau	4,471	506	(11)	12	4	(33)
Batu Brak	3,134	942	(30)	9	3	(33)
Suoh	12,326	3,914	(32)	10	5	(50)
Sekincau	9,423	2,317	(25)	9	3	(33)
Way Tenong	8,351	2,586	(31)	14	0	(0)
Sumber Jaya	8,908	2,973	(33)	14	0	(0)
TOTAL	82,057	22,848	(29)	169	42	(25)

Sources: Badan Koordinasi Keluarga Berencana Nasional [National Coordinating Office for Family Planning] 2001; Pembinaan Masyarakat Desa [Office for Village Empowerment] 2001.

A relatively high population density is another characteristic of the Sumber Jaya and Way Tenong Highland region. In 1961 the region had only 16,000 inhabitants, but the population rose to 25,000 in 1971 and then tripled to 75,000 by 1986 (Sevin 1989: 307). By 2000 the region was home to nearly 80,000 inhabitants living in 28 administrative villages. The dramatic increase in population, village development, agricultural production, and commercial activities during the last three decades has transformed the region into a lively

---

7 Notwithstanding that the IDT program failed to target the rural poor because many actually lived in non-IDT villages. Only through transforming the livelihoods of poor families in the IDT villages were benefits from the subsequent poverty alleviation and rural development programs derived (Perdana and Maxwell 2004).

area. What makes it exceptional is that the transformation of the region took place in the absence of large-scale government projects and private investment such as mining, plantations, or transmigration settlements.

During the 1970s and 1980s, the New Order village development program facilitated the creation of more administrative villages. Each village attracted development funds which had been made possible by the national oil boom and international lending institutions. This led to increased infrastructure development in the region. Creating more administrative villages was a justification to tap national development funds. This style of regional development became a central theme across all of Indonesia, particularly at all levels of government in Lampung Province, and caused more migration to the region.

The population in the region grew rapidly until the 1980s and then slowed. This was partly related to the 'closing down' of the state forest zones in the region. The late 1980s is remembered by the people in the region as the beginning of a number of efforts to remove smallholder farmers from state forest zones through a series of military operations, as well as the creation of a number of reforestation projects. The coffee boom that occurred during the monetary crisis was too short-lived to attract new migration.

The small rural towns of Sumber Jaya and Fajar Bulan and their immediate surroundings can perhaps be best characterised as a developing enclave in an underdeveloped district. West Lampung District has two other rural towns — the district capital of Liwa and the small beach town of Krui. The development of Liwa is largely due to its selection as the capital of the district in the early 1990s, bringing people and physical infrastructure to this otherwise quiet area. The rationale for the selection of Liwa as the district's capital was to separate the administrative centre from the commercial and economic centres. More development projects were carried out in Liwa after an earthquake in 1994 that caused extensive damage to the town as well as many other villages in Balik Bukit. The other town, Krui, used to be an important coastal trading centre for the west coast of Lampung in the colonial era. The people of Krui still believe that the reason their town was not selected as the capital of the district was primarily because of the high-profile lobbying efforts of a few powerful provincial bureaucrats and politicians who originated from Balik Bukit and Kenali.

Within West Lampung, Sumber Jaya and Fajar Bulan have had a very distinctive pathway of progress. Unlike Liwa, Sumber Jaya and Fajar Bulan were not selected by governments as key centres in the district. Unlike Krui, Sumber Jaya and Fajar Bulan are newly created population areas. However, the degree and level of modernisation in Sumber Jaya and Fajar Bulan is comparable to — if not surpassing — that of Krui and Liwa. Compared to other parts of this region, Sumber Jaya and Way Tenong are obviously more 'developed'.

In the wider context of the regional development of Lampung, it is important to note that the development witnessed in Sumber Jaya and Way Tenong is typical. Flourishing towns, many of which are bigger than Sumber Jaya and Fajar Bulan, can easily be found throughout other districts in the province. These towns include: Pringsewu, Gading Rejo and Gedong Tataan in the south; Metro, Bandar Jaya, Kota Gajah and Jepara in the centre; and Tulang Bawang in the north. All have been created mainly by Javanese transmigrants.

Sumber Jaya is the only designated receiving area for transmigrants from Java in West Lampung. The accommodation for these transmigrants in West Lampung as a whole is therefore much smaller than in other parts of the province.[8] A number of people in Sumber Jaya argue that it is partly due to the district not having many sites of transmigration that West Lampung still remains largely under-developed (*kurang berkembang*). Unlike other transmigration settlements located on the eastern Lampung plains and lowlands, where large-scale irrigation channels for rice fields can be built, Sumber Jaya is in a hilly mountain region where there are no large areas available to house such large-scale irrigated rice cultivation. Sumber Jaya and Way Tenong can provide anything anyone expects from modern rural Indonesia. In each of the small rural towns of Sumber Jaya and Fajar Bulan, in what the people simply refer to as the market, hundreds of shops and stalls are open seven days a week. There are also weekly rotational markets on Mondays in Fajar Bulan and Saturdays in Sumber Jaya. Due to a previous prohibition on Chinese opening businesses in rural areas in Indonesia, only a few shops are owned and operated by Chinese petty traders. In these shops people can get many kinds of goods including: food or meals; a variety of household goods such as cloth, electronic equipment and furniture; building materials; automotive spare parts; and brand new motorbikes. There used to be a movie theatre in Sumber Jaya but it no longer operates because of the influx of VCD players and pirated VCD rentals. Watching national dramas (*sinetron*), dubbed imported serials, and television news programs are the most common evening home entertainment.

Landline telephones, available in the nearby small town of Bukit Kemuning, had not yet reached the region in 2003. People used recently established cellular

---

8   Besides Sumber Jaya, Biha in Pesisir Selatan sub-district is another small-scale transmigration site in West Lampung. In the early 1990s, hundreds of forest squatter families from various parts of Lampung were resettled there under the local transmigration program.

phone services to communicate with relatives or colleagues nationwide, as well as occasionally to hear news from families working overseas (for example in Saudi Arabia or Malaysia).

Simply put, people in the region do not need to go to bigger towns or cities to get the goods and services they want. Unlike their parents, youths do not need to go to other towns to obtain a high school education. However, increasing expectations have accelerated the trend for people to travel out of the region. The desire to have a better or newer model of some consumption good, to take care of children's higher education, and to see the world outside of the region are the most commonly stated reasons for people to travel to bigger towns and cities within the province (Kota Bumi, Bandar Jaya, and Bandar Lampung) and in Java. Since the construction of the Western Sumatra Highway in the early 1990s, bus connections between Padang, Bengkulu, and Jakarta have made travelling to Java easy. Almost every week there are also special buses that travel from Sumber Jaya to Bandung. There are two types of buses: the cheap and popular *ekonomi* (non-air-conditioned), and the comfortable air-conditioned bus that promises to arrive on time. There are also minibus-taxis that pick passengers up at home in Sumber Jaya and drive them to any address in Bandung and the surrounding towns in West Java. Those who want to travel within the province usually take the buses that regularly travel from Krui and Liwa to the capital of the province, Bandar Lampung. When travelling in a group with families and relatives, a chartered car is the favourite choice. One can easily find a roadworthy vehicle to rent from a fellow villager. Celebrating Muslim holy days (*lebaran*) and attending the weddings of relatives are occasions where a chartered car is used. The flow of people from the region to and from cities in Lampung and Java not only blurs the rural–urban distinction, but also makes the distinction between Java and outer Java seem less relevant.

However, some within the region feel that there have been some negative consequences of being close to an urban centre, including increased criminal activity. For example, trucks and cars passing along the West Sumatra Highway often take rest stops at Fajar Bulan and Sumber Jaya where there are plenty of restaurants and food stalls with the popular Sunda and Padang menus. For overnight stops there are a number of small hotels with growing prostitution businesses. It is rumoured that there was once a romance stall, which beside food also provided young girls for men's sexual pleasure. The food stall soon became popular, especially among truck drivers. The local community, led by the religious leaders, soon took action. The stall owner was asked to stop the practice of prostitution and the girls were asked to leave, among them one from a neighbouring village. When asked if there was still prostitution in the region, the most likely answer was that 'there are none that provide the service openly'.

Another concern the villagers have is the use of drugs among the youth population. On one occasion, local policemen were suspicious that a small group of teenagers was using drugs at late-night gatherings in parking lots and bungalows constructed by the tourism office for sight-seeing and rest stops between Fajar Bulan and Sumber Jaya. On another occasion, a police officer caught and jailed a young man planting hundreds of cannabis plants in a capsicum chilli garden in one of the villages in Way Tenong. Security is another concern, and one which has led to the creation of night watches. Stories of brand new motorbikes being stolen are frequent, and burglaries are also frequently reported, especially during the coffee harvest season.

Within the region, people use motorbike taxis, minibuses, pickups, and four-wheel drive jeeps to get from village to village. Jeeps are only used on limited occasions such as to carry bulky materials from one rotational market village to another, to deliver heavy loads from the store or market to smaller stalls in hamlets in the hills, or to bring piles of dried coffee cherries and dried beans during the coffee harvest season down from the hills and mountains. A few jeeps can still be seen loading housing construction materials or transporting lumber from the remaining forests in the mountains. With more and more paved roads and bridges being constructed, the use of these off-road vehicles, which were very popular during the last three decades, has gradually declined.

The proximity of Sumber Jaya and Fajar Bulan — the respective capitals of two adjacent sub-districts separated by only a 15-minute drive — is a unique setting for upland rural Java. In other areas, the distance between the capitals of two neighbouring sub-districts typically takes an hour or more to travel. Sumber Jaya and Fajar Bulan, apart from being the primary places to sell local goods, are also where agricultural produce from surrounding villages is sold. A person travelling along the Western Sumatra Highway and viewing Sumber Jaya and Fajar Bulan might get the impression that the region is home to well-off rural Indonesians. Along this main road are modern brick houses and large traditional wooden stilted houses with either a motorbike or car in the front yard and a satellite dish on the roof.[9] A number of the houses have two storeys and are extremely luxurious. Indeed, most of the richest people in the region live in and near Fajar Bulan and Sumber Jaya and derive their wealth from the coffee trade and from retail shops. But the picture changes as one travels to the surrounding villages. Along the main road are compact settlements with rows of sturdy brick and wooden stilted houses, but as one goes farther from the main road and the main village settlement compounds, huts made of wood and bamboo start to fill the landscape. In the two town areas, many people are

---

9   Without the dish, only one of nearly ten national television channels could be received in the western half of the region and none in the eastern half.

involved in rural trading and other non-farm business and work, but in the surrounding villages and region the majority of people primarily derive their livelihood from small-scale agricultural production.

The Making (and Unmaking) of a Coffee Pot

Located on the eastern slopes of the Bukit Barisan mountain range, the villages in Sumber Jaya and Way Tenong region are surrounded by mountains and hills. In the centre is Bukit Rigis, to the north are Bukit Remas and Subhannallah, to the east are Gunung Abung and Bagelung, and to the west is Gunung Sekincau. The mountains are connected by gently rolling ridges encircling Bukit Rigis. The Way Besai River runs from Gunung Abung to the west, encircling Bukit Rigis and then down the valley to the west. At the easternmost end of the valley, at 720 metres above sea level, where the Way Besai River flows out of the region, is the site of the dam for the Way Besai hydroelectric power plant. Village settlements are located in the valley encircling Bukit Rigis on the banks of the Way Besai River.

Patches of forest can still be seen on the steep slopes and on the top of the mountains. Smallholder robusta coffee gardens are the predominant land use system, while wet rice fields are limited to the narrow banks of creeks and along the Way Besai River. All of the villages in the region have patches of wet rice fields, but villages with more than 100 hectares of these fields are rare. Rice is imported from other regions within the province and from Java. Within the settlements, many houses have a fish pond and favourite fishes such as goldfish (*ikan mas)* and gurame are regularly imported directly from towns in West Java such as Cirata, Cianjur, and Parung.

The dominance of smallholder coffee gardens in this particular region is a recent trend. Three decades previously, the region was heavily forested. While the expansion of wet rice fields and settlements has been limited, the transformation of primary and secondary forests into coffee gardens has been massive. Some consequences of this deforestation have been increased wild animal attacks and infestations. In 1997, men, women, some labourers on a reforestation project, and several farmers were attacked and killed by a tiger in Lebuay. The animal was later hunted down by a special team from the forestry office and brought to Taman Safari Zoo near Jakarta. Near the few remaining forests, villagers sometimes see tigers, bears, and deer, and the latter are still the object of non-commercial hunting. Monkeys, pigs, and elephants are now becoming pests, and attacks from pigs and elephants are especially serious. The local health clinics frequently receive patients that have been seriously wounded by pig attacks when the pigs are being hunted for destroying rice fields. Elephant groups that sometimes come to the villages seeking food during droughts have been another problem. Villagers have been forced to conduct extended patrols to keep the elephants away because of the limited number of forest rangers. Local people think that killing the elephants would be the easiest way to protect themselves, but the fear of jail for killing endangered animals generally stops them from doing so.

The transformation of forests into smallholder coffee gardens has been accompanied by a decline of livestock husbandry in the area. Elders confirm that in the 1980s, the old Semendo villages were full of cows and buffaloes. Now only a few households in each village rear these animals. As a result of the expansion of coffee gardens, neither grazing land nor labour to feed the livestock are available. There is a possibility that this trend began after the confiscation of cows and buffaloes during the Japanese occupation in 1942–45. Tiger attacks were the primary reason for the previous reduction in sheep and goat numbers. Until recently, village night patrols had to be conducted in some of the villages to prevent tigers from taking the sheep or goats from the stalls. With the further shrinking of their habitat, the tiger population seems to be gradually declining, and more sheep and goats are now seen in the region.

With no forest left near the villages, another difficulty now is how to obtain timber for housing. Favourite first class timber from the forests, such as *tenam*, *cempaka*, and *medang*, has become very expensive. In the 1990s, the price of such timber was equal to local costs of cutting and transport, but the price has become more than double those costs. Cheaper timber from plantation trees is now preferred, and shorea and exotic *afrika* are now used for housing construction and furniture. Shorea and teak imported from the nearby regions are now sold in local lumber shops. Inferior quality timber such as *kapuk* and *dadap* are also used for light construction, such as huts and kitchens attached to the main house. While the conversion of forest to smallholder coffee gardens is obviously one cause of the scarcity of local timber, illegal logging has been another important factor. In most villages, some of the village elites have engaged — and in some cases continue to engage — in this lucrative yet illegal business with the backing of the police, military, or forestry personnel.

A large part of the region is gazetted as state forest reserve and mostly classified as protection forest. To the west there is Bukit Barisan Selatan National Park. Local people have called these zones 'state forest land' when referring to the land and 'state forest' when referring to the forest. During the 1980s and 1990s, the region was home to forest protection and rehabilitation projects. Yet there is no evidence to show that efforts to convert present coffee stands into plantation forests and prevent further expansion of smallholding coffee farming have been successful. On the contrary, plantation forests have been transformed back to coffee gardens, with the remaining natural cover also continuing to be converted.

Table 3-3: Area, population, population density, and poor families in Sumber Jaya and Way Tenong sub-districts.

Village	Area (km²)	Persons per km² (1998)	All families (2000)	Per cent poor families (2000)
Simpang Sari	44	195	1,424	30
Sukapura	9	286	437	32
Way Petai	26	154	668	38
Suka Jaya	18	100	485	28
Sindang Pagar	143	13	380	43
Tri Budi Sukur	10	213	635	24
Pura Jaya	15	233	751	26
Purawiwitan	10	245	657	28
Muara Jaya I	9	131	388	43
Muara Jaya II	9	190	455	36
Pura Mekar	17	269	1,034	31
Gedung Surian	14	140	526	44
Cipta Waras	15	111	458	43
Tri Mulyo	15	173	610	39
SUMBER JAYA TOTAL	356	115	8,908	33
Pajar Bulan	17	282	1,238	17
Puralaksana	22	160	691	22
Karang Agung	19	157	627	31
Mutar Alam	9	499	498	36
Sumber Alam	5	261	703	31
Tambak Jaya	10	302	512	40
Tanjung Raya	29	86	755	22
Sukananti	5	866	683	28
Sri Menanti	7	130	210	58
Sukaraja	29	79	442	40
Padang Tambak	12	285	614	34
Sidodadi	5	343	387	47
Semarang Jaya	5	248	366	48
Gunung Terang	20	108	625	34
WAY TENONG TOTAL	193	200	8,351	31

Source: Website of the West Lampung District Government (viewed 2004).

The region of Sumber Jaya and Way Tenong is recognised as an important 'coffee pot' within the province. The region can perhaps be regarded as the most intensive smallholding coffee growing area in the province due to recent cultivation practices. All techniques and inputs have been applied to achieve maximum output from coffee farming. Grafting, where tissues of different coffee varieties are joined together, has been done since the early 1990s. Initially, twigs of more productive varieties of robusta were brought from the nearby region of Tanjung Raya where a handful of farmers had successfully obtained higher production after grafting their old coffee trees with stock imported from Jember in East Java. Chemical fertilisers have been used since the late 1970s when they were heavily subsidised and made available under the New Order's famous Bimas and Inmas scheme intended for rice cultivation. Local traders, usually wealthy villagers, created schemes exploiting familial ties to obtain delivery orders from a designated fertiliser wholesaler to purchase fertiliser individually or in a group. Stalls and stores would often also sell chemical fertilisers. Various techniques of soil conservation, such as the use of terraces, ridges, and pits are applied as well, resulting in increased production levels. During 'normal' years the average production in the region is 1,000 to 2,000 kilograms per hectare — much higher than the national average of coffee production, which is around 500 kilograms per hectare. Only during 'poor' years does the production in the region fall to around the national average.

The cycle of 'good', 'normal', or 'bad' years is perceived to be the result of the interplay between coffee prices, climate, and the age of the coffee gardens. The late 1950s, late 1980s, early 1990s, and the end of the 1990s — the *krismon* period of 1997–98 — were considered to have been 'good' years. The 1980s and the period from 1999 to 2002 were considered to have been 'bad' years and the remaining years were considered 'normal'.

The price of coffee is considered to be good or bad when in comparison to the price of basic necessities, most importantly milled rice. For example, during the 'good' years of the *krismon* in 1997–98, a kilogram of coffee was selling for Rp 8,000–12,000 and a kilogram of rice for Rp 500–1,000. From 1999 to 2002, the price of coffee dropped to Rp 3,000–4,000 per kilogram while the price of rice rose steeply to Rp 2,000 per kilogram. During these 'bad' years, a kilogram of coffee was almost equal to the price of a kilogram of rice. To make matters worse, the price of other goods also rose.

The 1950s was said to have been the beginning of the 'good' years as far as coffee farming was concerned. A kilogram of coffee was selling for Rp 3.5 while four kilograms of rice was said to cost only Rp 1 in this region. The late 1950s was also said to be a time when the practice of transforming upland rice swidden into coffee garden on fallow land ended. More labour was hired and a day's work earned Rp 3.5, equal to the price of one kilogram of coffee. Rather than being left fallow, old coffee gardens were kept in production. More Javanese and Sundanese arrived, either as labourers or sharecroppers or, for those who had some capital, to buy young and old gardens and abandoned or fallow fields. Old gardens were pruned and rejuvenated. New forests were cleared and planted with upland rice for one or two crops while also being planted with coffee. Transforming the cleared forest into coffee gardens, without the early stage of swidden, was also commonly practised. Opening several plots of different ages was necessary to ensure continuous production during the 'good' years, with gardens of full bearing coffee trees aged three to seven years. Pruned and rejuvenated gardens produced a relatively constant annual production, although lower than during the 'good' years. As a diversification strategy, the traditional system of inter-planting coffee with pepper continued to be practised by some farmers in the region. Besides providing shade for the coffee, *dadap* and *gamal* trees functioned as the poles for the pepper vines. More recently, commercial tree crops (for example, timber and fruit) have also been planted in coffee gardens.

Despite the introduction of chemical fertiliser, the late 1980s were considered to be bad years for coffee. Cocoa and cloves gained in popularity as a crop substitute. Many coffee gardens were transformed into either cocoa or clove gardens. Cocoa grew and produced well, but there was no one to buy the harvest, while the cloves were almost completely destroyed by leaf blight disease. A few clove trees still survive, but their economic importance in the region is insignificant. Coffee, however, has never disappeared, and the failure of both cocoa and cloves brought smallholders back to coffee.

The 1980s and 1990s are remembered as the decades when government agricultural extension programs came to the villages. New techniques and new inputs were introduced. Smallholders were encouraged to form farmers' groups, with whom field extension officers worked closely to develop demonstration plots for better farming techniques. A World Bank-sponsored program to boost Indonesia's smallholder export crop production, Proyek Rehabilitasi Tanaman Ekspor (Export Crops Rehabilitation Project), provided cheap credit for replanting and chemical fertilisers for hundreds of hectares of coffee gardens in the region. The forestry office ran projects to introduce soil conservation techniques (terracing and tree planting), also on the demonstration plot basis. The coffee exporters' association (Asosiasi Eksportir Kopi Indonesia) regularly provided grants, both directly to farmers' groups and through agricultural extension agencies, to deliver technical assistance to promote better quality coffee production. Sponsoring farmers' delegates to visit and learn from other coffee pots in Java was one form of technical assistance.

The 1990s was the period when the harvesting of coffee enabled the people in the region to secure a higher economic position. Many brick houses were built during the first half of the decade. Old traditional stilted houses were renovated and new ones constructed. Cars and motorbikes became much more numerous. Local coffee traders got richer and petty trading flourished. The prohibition preventing the Chinese from opening businesses in rural Indonesia enabled a few merchants in the region to accumulate considerable wealth from local commercial activities. The climax came during the nation's monetary crisis when the coffee price skyrocketed. Farmers received export dollars for their crops as the value of the rupiah deflated. The El Niño drought brought good production from mostly grafted coffee trees and the price of coffee rose three to four times while the price of other goods remained stable. With the sudden increase of purchasing power, local people likened the massive buying of luxury goods such as cars, televisions, and furniture to buying cheap snacks: 'Just like buying fried bananas!'

It was also during the 1990s that the dwarf arabica variety was introduced, again on a demonstration plot basis. Seedlings were distributed free of charge and farmers were told that arabica would sell for a higher price, but that never happened. Local traders and exporters bought both robusta and arabica at the same price and, according to those who planted arabica, more labour was required to maintain their gardens, especially to remove the twigs. Additionally, unlike robusta, without chemical fertiliser the arabica would bear no cherries. These factors all contributed to the lack of conversion from robusta to arabica in the region.

The post-*krismon* economic recovery of Indonesia beginning in 1999 brought a real economic crisis for the villagers in the region. The price of coffee dropped dramatically while the price of rice and other basic goods rose steeply. Things became very difficult, and even buying cheap fried bananas was no longer easy. Too much rain was blamed for the drop in average production in the region's coffee gardens as well as the change in the use of chemical fertilisers. Some simply said that the coffee trees were exhausted after the long 'good' years in the 1990s.

While the 'bad' years of the 1980s drove some smallholders to cocoa and cloves, some of the smallholders in the region began to turn to commercial vegetables. Vegetable production in Liwa declined due to a combination of vegetable fields being converted into coffee gardens during the *krismon*, a recent severe disease infestation, and a decline in soil fertility. However, a steadily expanding vegetable production in the neighbouring region of Sekincau, to the west, inspired the conversion of some coffee gardens into vegetable fields and the interplanting of coffee and small hot chilli throughout Way Tenong and Sumber Jaya. In 2002, in the towns of Fajar Bulan and Sumber Jaya, one could hardly miss seeing sacks and baskets of vegetables filling the storehouses and being loaded onto pickups or light trucks for export to larger provincial towns and sometimes to Java.

A Multi-Ethnic Middle Peasantry

The slower pace of migration into the region since the late 1980s has influenced the current pattern of landholding in the region. It has helped prevent the further monopoly of land by a select few and further increases in landlessness.

As far as landholding is presently concerned, the region has not evolved into polarised and opposed classes, with a few landlords at one end and a mass of landless people at the other (see Table 3-3). This is mostly due to the domination of a middle peasantry in the region, which is not simply a function of the land-to-person ratio or population pressure alone, but is linked to wider and more complex processes.

'Wealthy' landholding in the region refers to ownership of more than 10 hectares of land. It is everyone's dream to have such a large amount of land, but few are able to do so. In almost all of the villages, only a small number of families with around 10 hectares of coffee gardens can be found, and someone owning more than 20 hectares has 'never been heard of'. There are two strategies for acquiring a large garden. One is by organising a group of men for forest clearing. The leader is responsible for recruiting and providing the food for his working party during the forest clearing and coffee planting, and retains a larger portion of the newly established gardens. No cash payments are involved; instead each group member receives a portion — a hectare or two — of the new garden, and then they have the option to sell or keep it. Some of the plots are sold to recover costs, such as providing food for the working party. The members can plant upland rice on the newly cleared land for one or two crops and are entitled to all of the harvest. A second strategy is to acquire the gardens during 'poor' years when their owners are in financial difficulty and when the price of the garden can be bought below the former market price. After a decade or two of following either strategy repeatedly, one can eventually own a large number of coffee gardens scattered throughout the area.

It is also important to note that these large garden holdings soon become fragmented and passed on to descendents. With the fluctuation in coffee prices and production during 'poor' years, the revenue from coffee alone is insufficient to cover the cost of its upkeep, including fertiliser and hired labour. Having the plots scattered over large areas and with coffee trees at different stages of development makes supervision difficult and production uncertain. Therefore it is necessary for large landowner families to have sources of income other than their coffee gardens, such as rice fields, trading and transportation activities, and sometimes money lending.

Having a large amount of land has been discouraged by the national legal system. The regulations dictated by the *Indonesian Basic Agrarian Law* of 1960 set limits on the plot size of land that can be individually owned.[10] Beyond the set limit, the owner can only obtain a long-term lease on a plot of land which is time

---

10   Government Regulation 56 stipulates a 5-hectare ceiling for irrigated land or 6 hectares for non-irrigated land per family in areas where population density exceeds 400 persons per square kilometre. For areas with less than 51 persons per square kilometre, the limits are between 15 and 20 hectares.

consuming and incurs considerable cash payments. More importantly, leasing land is incompatible with the traditional system of inheritance that emphasises land ownership with or without an official certificate of title. Certificates of land ownership, on the other hand, are easier to obtain and are much cheaper under the government land administration projects regularly initiated in the region. There are several cultural motives for selling land, including the ability to invest in more profitable or less risky businesses and to obtain cash for various uses, such as children's higher education and marriage, treatment of severe illness, completing construction of a home, or, less commonly, for pilgrimage costs to pay homage in Mecca.

Landless and near landless farmers are not uncommon. They are latecomers and/or spontaneous migrants who have settled in the region as labourers or sharecroppers. Young couples waiting to inherit land from their parents also fall into this category. Finding a garden and/or rice field to sharecrop is not difficult in the region. Borrowing an 'unused' plot without paying is another arrangement to which landless households resort in order to gain access to land for cultivation. Villagers in this stratum often earn income from working in gardens that belong to their friends and neighbours. Going from landless labourer to smallholder is a common form of upward mobility that utilises the popular tactic of saving money during the 'good' years and using it to buy land. The bulk of the population in the region owns one or more plots totalling at least one hectare of coffee garden (see Table 3-4). To maintain more than a hectare of coffee garden requires extra labour in addition to that of household members. This necessitates the use of previously unused plots (*numpang*), sharecropping, and hired labour.

Table 3-4 shows the distribution of coffee gardens and rice fields between a sample of 107 households from seven hamlets (and seven villages) in the Sumber Jaya, and Way Tenong sub-districts. The villages selected for the survey represented old Semendo villages created prior to the 1950s (Gunung Terang and Sindang Pagar), transmigration villages created in the 1950s (Simpang Sari and Fajar Bulan), and newer villages created by the subsequent spontaneous transmigrants since the 1960s (Cipta Waras, Suka Jaya, and Trimulyo). Hamlets with rice fields from each village were chosen for survey in consultation with village leaders. About 20 per cent of the hamlet residents were chosen for the household survey. The survey excluded hamlets without rice fields and non-landowning households — namely sharecroppers and/or contract labourers — many of whom lived in houses or huts in the gardens outside the hamlet settlement compounds.

Table 3-4: Land ownership in selected hamlets in Sumber Jaya and Way Tenong sub-districts.

Village	Hamlet	Sample h'holds	Coffee Gardens				Rice Fields		
			% owning	Size range (ha)	Av. size (ha)	% owning	Size range (ha)	Av. size (ha)
Gunung Terang	Gunung Terang	16	100	0.5–4.0	1.6	31	0.5–1.75	1.0
Sindang Pagar	Sindang Pagar	13	100	0.25–6.0	2.4	46	0.25–1.0	0.6
Simpang Sari	Air Ringkih	14	85	1.0–3.0	1.25	50	0.04–0.25	0.09
Fajar Bulan	Fajar Bulan	12	91	0.25–3.0	1.3	43	0.25–1.0	0.5
Cipta Waras	Waras Sakti	18	100	0.4–2.4	1.3	35	0.2–1.0	0.5
Suka Jaya	Talang Bodong	15	100	0.5–4.0	2.0	43	0.04–0.75	0.6
Trimulyo	Air Dingin	19	100	0.25–12.0	2.6	15	0.25–0.25	0.25

Source: 2002 survey data.

Engaging in various forms of off-farm work is a general strategy among all strata though the motivations, processes, and consequences may differ. Among the lower economic stratum, since income is insufficient, survival is a primary goal. About one in three households/families in the region were classified as poor in 2000 (see Tables 3-1 and 3-2). Family member(s) were sent outside of the region to work in cities in Java or ideally in foreign countries. For families in the upper stratum, investing in more profitable and less risky businesses was a primary goal. For all strata, having educated children who will no longer need to engage in farming was a worthy goal.

In the Sumber Jaya and Way Tenong region, illiteracy is relatively low, especially among the younger generation. Most elders and adults have received a primary school education in the region, and it is common now for the younger generation to continue on to junior and senior high school. Among the lower stratum, however, money is a large constraint that prevents children from getting a higher education. Money is also a concern among the middle stratum, though not to the same degree. For the upper stratum, it is the children's desires that determine how far they pursue their education. Among the middle and upper classes there are many cases where children are reluctant to undertake further studies or incapable of doing so. The children's reasons for not continuing with schooling are accepted and justified by parents as a growing number of those graduating from universities take low-paid jobs or fail to find a job altogether. In these cases, studying at university is considered a waste of time and money.

Higher education and socio-economic mobility are possible partly because of the acceptance of the government family planning program, Keluarga Berencana (KB). The majority of fertile couples in the region participate in the family planning program. Previously subsidised, fertile couples now pay for the KB injections and pills that prevent pregnancy. On one hand, having fewer children increases the ability of parents to financially support their children's education, but on the other, it reduces the availability of free labour for farming, which again necessitates the use of sharecropping and wage labour to make ends meet.

Children's education and home construction/improvement are two priorities, and income generated in excess of everyday household needs goes towards paying for these two items. Buying a vehicle and household equipment are the next priorities, and the last household financial demand is to 'take a last step to the stairway to heaven', or a pilgrimage to Mecca. There are two types of pilgrimages (*haji*). The first is called *kiyai haji* and the pilgrim is bestowed the title of 'real *haj*' (*haji betul*). This pilgrimage is made by those with a deep knowledge of Islam, who apply it in daily life, and who are actively teaching Islamic religion to *pesantren* pupils — students at an Islamic school — in mosques or occasional learning groups. There are few with this status and they earn high respect. The second type is referred to as 'coffee *haj*' (*haji kopi*). These individuals were able to make the pilgrimage to Mecca with the earnings of their large coffee garden holdings. Their knowledge of Islam and the alignment of their daily life with the teaching of Islam are limited. Compared to the *haji betul*, the *haji kopi* are more numerous, while Semendonese and Sundanese *haj* are more numerous than Javanese *haj*.

Distinguishing the proportions of the three major ethnic groups in the region is difficult. None can be said to be dominant. In the village markets, apart from Bahasa Indonesia, all three languages — Sundanese, Javanese, and Semendonese — are spoken interchangeably. The younger generation usually understands all languages and most speak all three. Since there is neither an ethnic preference nor avoidance in marriage, intermarriage is prevalent. With marriage, it is religion that will determine compatibility. As long as the religious denomination and level of devotion is the same, inter-ethnic marriage is acceptable.

Within a village it is common to find a hamlet or neighbourhood with a dominant ethnic group — Sunda, Java, or Semendo. Those from other ethnic groups living in a hamlet adopt the dominant language. There are also hamlets and neighbourhoods with a more diverse mix of ethnic groups along the main road and Bahasa is spoken here. Along the main road in the main village settlements, Padang traders and tailors and Batak tyre repair services are common.

It is important to note that with regard to identity, all of the migrants from the highlands of Palembang see themselves as Semendo although they may originally have come from other Pasemah sub-groups. Thus, all Pasemah-speaking persons in the region identify themselves and are identified as Semendo. The same is true of those from Sunda, including those few from Banten who identify themselves and are identified by others as Sundanese.

Ethnicity in the region is often a subject of political mockery. In the case of forest destruction, the migrants from Java 'wash their hands' of this issue and instead blame the Semendo for their aggressive yet admired techniques in clearing the forest. In retort, the Semendo point out that it is the migrants from Java who farm the cleared forestland. The Semendo claim that the migrants from Java have only become as 'healthy' as they are now thanks to Semendonese generosity in 'giving' them land. The migrants from Java claim that the region's progress is the result of their work, and that without them there would be no development or progress. These friendly rivalries over the subject of development and progress provide the central and dynamic theme of local village politics. The next chapter will discuss this dynamic as it relates to development and progress in the region.

4. Local Politics: Bringing the State to the Village

Sumber Jaya and Way Tenong have been the targets of constant national, regional, and local political manoeuvering to control its population. There are clear indications of deep state penetration into the villages. Local people are increasing their efforts, through their village leaders, to expand state participation in the village as a strategy to tap state resources and put their village in the mainstream of national and regional politics. These processes have led to the emergence of politically powerful village elites whose power is still both limited and circumvented due to villagers' ability to develop procedures that constrain the emergence of individuals with dominant political power in the village.

Military Campaigns against State Enemies

From the mid 1960s to the late 1980s, villagers in Sumber Jaya and Way Tenong experienced multiple military operations designed to crush rural dissent. A military operation to wipe out the communist movement occurred in the mid 1960s, and another operation against religious rebels happened in the late 1970s. These actions created a dynamic relationship between the villagers in the region and the modern state.

Chasing the Communists

During the military campaign against Indonesia's communist party, Partai Komunis Indonesia (PKI) and its elements in the mid- 1960s, hundreds of men and women were taken from their homes, loaded into trucks, and jailed at the military post (*koramil*) in Sumber Jaya for interrogation. Some of them were taken to other military camps in Kotabumi, and some of them never came back. Some spent years in jail and the rest — the majority — returned to the *koramil* at Sumber Jaya. During the following years these women and men were obliged to report regularly *(wajib lapor)* to the *koramil* and were treated as corvée labour (*kerja bhakti*) repairing roads and cleaning military, police, and public facilities. The sight of hundreds of men and women carrying their children in fear and sitting in the sun in front of the *koramil* office and enduring various forms of torture and intimidation has filled the memories of many people in the region.

The alleged Communists came from almost all corners of the region, but the largest proportion were said to be from Simpang Sari and Way Petai. However, it was later revealed that the majority of these Communists had not engaged in any meaningful political action. In the region, the PKI never gained a significant number of votes during the early national elections. In 1965, prior to the commencement of the national military campaign against the PKI, women were recruited to join various Islamic teaching groups (*pengajian*) and cottage industries (for example, sewing or stitching), and young people were encouraged to join the *rebana* (tambourine) religious music groups. The only indication of concrete action, it was said, was regarding land reform, and it was rumoured that landless villagers were organised into groups in anticipation of obtaining ownership of farming land. Threatened with becoming the targets of dispossession, village elites and large landowners were more than willing to give full cooperation to the military personnel.

During the campaign there were stories of villagers mistakenly detained (*salah tangkap*), and villagers with no links whatsoever to the PKI were interrogated and subject to intimidation by the *koramil* personnel. This was largely the result of fierce opposition between factions competing for power in the village. Both sides gave information on their opponent's involvement with the PKI. Having a distant relative or friends involved in the PKI movement was enough to bring someone to the notice of the *koramil*.

Suspicion of involvement in the PKI had long-term deleterious consequences for some. Near the market town of Sumber Jaya there is a small hamlet, many of whose inhabitants were the victims of oppression during the anti-communist campaign. Until recently, the hamlet has been isolated, receiving no government projects that neighbouring hamlets received, such as roads and schools. Most of its poor inhabitants have lived mainly as labourers and sharecroppers, or by tree felling and cutting from the remaining forests nearby.

Chasing the Islamic Rebels

While no 'concrete action' by the communist movement ever occurred, an Islamic rebellion a decade later had different results. Warman and his *gerombolan* (group or band of men) were remembered as having a strong anti-state agenda and multiple criminal records. In the second half of the 1970s, Warman and his followers were involved in some armed encounters in various parts of north Lampung. The *gerombolan* were responsible for burglaries, raids on buses, killing village officials, and attacks on military posts from which the group obtained firearms. The last two activities were said to have been more frequent during the New Order's 1977 national election, and were widely perceived as an attempt to sabotage that election.

Warman was believed to have been one of the staunchest followers of Kartosuwiryo, the leader of the Darul Islam (DI) and Tentara Islam Indonesia

(TII) movements that were founded in 1949 in West Java. The ultimate political agenda of DI/TII was an Islamic state. After more than a decade of warfare with the Indonesian army, the DI/TII rebellion was crushed and Kartosuwiryo was executed in West Java in 1962. Warman fled to Way Tuba, a region near the town of Baturaja in the neighbouring province of South Sumatra (Palembang). In 1975–76, he and his family moved to Sukapura in Sumber Jaya. About 50 of Kartosuwiryo's followers joined the BRN transmigration in the 1950s and lived in Sukapura. Of these, about 15 to 20 later joined Warman. During these years, none of his neighbours knew that the notorious Warman was living next door or that their village was the headquarters of his *gerombolan* movement. Warman led a *pengajian* (Qur'an reading group) in his small *mushala* (praying house). A type of 'true Islam' (*Islam sejati*) was Warman's main political teaching, and when the group became more and more exclusive and held separate Friday prayers instead of attending the village mosque, the village officials and military began to investigate. Soon the hilly region of Sumber Jaya and Way Tenong became a battleground between the *gerombolan* and the military troops.

Instead of surrendering to the military troops, the *gerombolan*, consisting of no more than 60 men, fought back relentlessly. Hiding in the forest during the day, they raided military posts and villages in the night. As in the DI/TII movement in West Java, food supplies were taken from shops and stalls (*warung*) belonging to villagers. Unlike the DI/TII rebellion, the local *gerombolan* did not terrorise the whole village, apart from taking food from the *warung*, and targeted only village officials. In fact, it was military personnel that forced ordinary villagers to take part in the campaign against the *gerombolan*. However, villagers were not allowed to carry firearms, providing them with an excellent excuse to avoid becoming involved in warfare against the *gerombolan*. Therefore, casualties were limited to *gerombolan* members, military personnel, and village security officers (*hansip*). Although most of his followers were shot dead or captured, Warman himself escaped, first to another location in Lampung and then to Java. The military hunt for Warman continued, and Ketapang, near Kotabumi, was the site of a fierce clash between the *gerombolan* and military troops resulting in fatalities on both sides. The battle was commemorated with the building of a *koramil* post.

After Warman fled to Java, he was captured in Magelang but managed to escape and remain at large until 1978 when a team of *Kopassus* (army special forces) shot him dead in Soreang, near Bandung, West Java. Like the victims of the military action against the PKI, a few surviving members of Warman's rebellion and the wives and children of those who died or were jailed now live in isolation and poverty. Many moved elsewhere in Sumatra or across to Java.

The relatively long period of the military hunt, the fact that the group of rebels was small, and the absence of casualties among ordinary villagers, all indicate that villagers in the region carefully positioned themselves in the battle. Ordinary villagers neither harboured the rebels nor fully assisted in the military campaign. Nonetheless, the alleged PKI movement and Warman's *gerombolan* rebellion in the region brought further state intervention to villages in the region which I shall now discuss.

National Politics in the Villages

Following the successful crack down on communist and religious dissent, a strong military presence continued in the region. The military's role expanded from hunting down state enemies to ensuring *monoloyalitas* (single or undivided loyalty) of the region's population towards the state. 'The state', until the 1998 *reformasi*, meant Suharto's New Order and Golkar.[1] At the heart of the New Order were the twin objectives of 'political stability' and 'development'. Both *koramil* officers and the *babinsa* (village military officers) played a key role in the process. To become the head of village *(kepala desa)* or to hold other official positions in the village, a clearance from *koramil* was needed in addition to the 'blessing' from the sub-district head (*camat*) and Golkar functionaries. Through a program known as ABRI Masuk Desa (AMD), which literally means 'the military enters the village', the villagers were forced to participate in *gotong royong* or *kerja bhakti* (community works) on village projects such as building and maintaining roads, bridges, and schools. Even in the absence of AMD, the constant supervision by village military personnel (*babinsa)* ensured villagers' participation in routine community works in similar projects, especially on the construction and upkeep of roads.

The triumph of Golkar until the 1999 national election, and the instalment of Golkar cadres in village administration, ensured a state of 'political stability' in the region. Undivided loyalty (*monoloyalitas*) toward the state was achieved by appointing village leaders to official positions in village administration, such as village social boards (*lembaga sosial desa*) and village boards for community resilience (*lembaga ketahanan masyarakat desa*), youth associations (*karang taruna*), mosque boards for religious leaders, and organisations devoted to family welfare education for women (*pendidikan kesejahteraan keluarga*).

The creation and incorporation of village leaders into the village administration was directly related to success in the mobilisation of rural populations in centrally planned rural development projects. In this region — as elsewhere in the nation — rural development projects included the construction of physical infrastructure (roads, bridges, schools, village halls, markets), village

---

1 The name Golkar derives from *golongan karya* (functional groups).

administration *(pemerintahan desa)*, expenditures such as transportation costs for village officials, economic development (for example, agricultural extension and land administration), and social welfare (family planning for example). The New Order agenda of political stability and development was successfully achieved in Way Tenong and Sumber Jaya. Due to the absence of villagers' political alignment with any group other than Golkar, the villagers in the region devoted themselves to the rural development agenda. It was during this period of political stability and rural development from the late 1970s to the mid- 1980s that more administrative villages were created and more people migrated and settled in the region. The mysterious nationwide killings of criminals in the early 1980s *(penembak misterius)* further 'stabilised' the region and enabled the movement of more people into it.

The political texture of Sumber Jaya and Way Tenong is a reflection of the political dynamic at the national level. With Golkar loyalists accounting for the majority of the population during the three decades of Suharto's New Order regime, the region received a share of the 'development cake' that was envied by the neighbouring regions. All villages have paved or gravel roads and there are at least two elementary schools. In every three or four villages there is a health clinic, rotational market, and junior high school *(sekolah menengah pertama)*. After the *reformasi* of 1998, local people in the region — like many people nationwide — switched their political loyalties to the previously suppressed Partai Demokrasi Indonesia Perjuangan (PDIP), not because of its political agenda, but simply because they had had enough of Suharto's New Order.

Winning the 1999 election had a very different meaning for the local PDIP functionaries. It was just like night turning into day. Economically and politically marginalised because of their deep devotion to Megawati, the 1999 election provide them with a harvest to reap. Party functionaries from Sumber Jaya played dominant roles in the PDIP's district branch, the district house of representatives or *dewan perwakilan rakyat daerah* (DPRD), and the administration of West Lampung. The positions of chairperson of PDIP, chairperson of the DPRD, and vice-regent of the district *(wakil bupati)* were all given to PDIP politicians from Sumber Jaya. Sumber Jaya and Way Tenong were also home to key figures from 'Islamic' parties such as PPP (Partai Persatuan Pembangunan*)*, PAN (Partai Amanat Nasional), PKB (Partai Kebangkitan Bangsa), PBB (Partai Bulan Bintang), as well as the former ruling party, Golkar. People in the region noted that the new members of the DPRD busied themselves with renovating their houses or building new ones and getting cars. This drastic change was most noticeable among many of those who were not previously among the well-to-do people in their villages.

What brought politicians from the region to the top seats of the district-level political arena was the sheer number of their voters. By the end of the 1990s, the two sub-districts of Sumber Jaya and Way Tenong were home to a quarter of the total population of West Lampung District. In 2002, West Lampung had nearly 400,000 people spread over fourteen sub-districts. Thanks to the high population numbers, Sumber Jaya and Way Tenong have always been seen as two important sub-districts in West Lampung. An important pocket for Golkar during the New Order, the region turned into the centre of PDIP and the 'middle axis' parties, including the PPP led by the vice-president Hamzah Haz, the PKB led by former president Gus Dur, the PAN led by Amin Rais, the chairman of the People's Consultative Assembly, and the PBB led by Yusril Ihza Mahendra, the Minister for Justice during the *reformasi* era. So in 2002 and 2003, an even newer village strategy was created. As some villagers in the region put it, 'we have to join the crowd otherwise we will be left behind'.

During 2002 and 2003, there were early signs of an alignment of the region's population to the established political parties, which were now the ruling party (PDIP) and the 'middle axis' camp. The national configuration of politics towards the national election in 2004 was also reflected in the region, as exemplified by the splitting of the PPP into a camp led by the vice-president Hamzah Haz and another camp, the PPP Reformasi, led by the popular Islamic preacher Zainuddin MZ. On one occasion, over a thousand people gathered on the Fajar Bulan soccer field to hear a speech by Zainuddin MZ inaugurating the branch of his PPP Reformasi in West Lampung as though the support from the region's population was assured. A couple of months later, brand new billboards supporting Hamzah Haz's PPP were erected in some villages, indicating that the village functionaries were active in getting local people's support. Similarly, when the PKB split, boards and banners of both factions (for and against Gus Dur) could be found throughout the region. In the market towns of Fajar Bulan and Sumber Jaya, one would see boards and banners of different political parties erected side by side. Only in Golkar did loyal cadres wait until the national election was closer before erecting billboards and banners. Politics in the region continued to reflect national political dynamics.

Sumber Jaya and Way Tenong have also been the location for mass organisations based on ethnicity and regionalism. In 2002, a branch of Paku Banten was inaugurated in Sumber Jaya and in the following months, a Batanghari Sembilan branch opening was celebrated in Way Tenong. The Paku Banten was formally declared to be an umbrella of all camps of *pencak silat* (martial arts) in Lampung. Paku Banten is known for its involvement in gathering mass support *(dukungan massa)* for particular candidates in the election of district heads *(bupati)* in the province. The most favoured candidates were already incumbents who hoped to be re-elected by the DPRD for the next term. The gatherings were organised with a *pencak silat* performance, *dangdut* (reggae) music entertainment, and concluded with a speech in favour of the candidates. Paku Banten is open to people of any ethnic background, but in Sumber Jaya, Paku Banten members and functionaries were Sundanese and Javanese, and many of them hardly practised *pencak silat*. Batanghari Sembilan was also officially formed as a venue for promoting the arts (singing and *pantun* poetry composition) of people originating from the southern part of Sumatra, including Jambi, Palembang and Bengkulu, but excluding indigenous Lampung people. Two national figures, Taufik Kemas (President Megawati's husband and a key figure in PDIP) and Ali Marwan Hanan (one of the chairpersons of PPP and the Minister of Cooperative and Small Business) were said to be involved in Batanghari Sembilan. In Way Tenong, Batanghari Sembilan functionaries are Semendonese politicians, government officials and businessmen.

In these mass organisations, the candidates for political positions will typically promise to bring 'progress' and 'development' to the region in exchange for the support of the region's population. These statements are what the people in the region are eager to hear to ensure that they will not be 'left behind.' Many see both Paku Banten and Batanghari Sembilan as a response to these mass organisations of the native Lampung population which, besides promoting Lampung arts and culture, also campaign for the filling of political positions by 'native children' (*putra daerah*). While the 'native children' have joined Paku Banten, none seem to have joined Batanghari Sembilan. The functionaries and prominent members of the mass organisations were key members and participants of other mass organisations during the New Order, such as Pemuda Pancasila, Angkatan Muda Pembaharuan Indonesia (Youth for the Renewal of Indonesia), Komite National Pemuda Indonesia (National Youth Committee of Indonesia), and the like. Formerly loyal to the state as their central theme, the groups now promote regionalism, but underneath is ultimately the struggle for local, regional, and national power.

Village Head Elections

By integrating their villages into the state, the villagers are involved in an effort to tap state resources to bring 'progress' to their villages and enable them to maintain their livelihoods and pursue prosperity. At the local level, state attempts to control the rural population and villagers' efforts to tap state resources are clearly visible. These dynamics repeatedly occurred during the New Order period as well as the period immediately after Suharto's fall in 1998.

In village head elections during the New Order, one way to position a Golkar functionary as the village head was by blocking the non-Golkar candidate's eligibility to obtain approval and letters of 'clearance' from the sub-district office. To ensure the victory, village head elections were often organised with a single favoured candidate against an empty box (*kotak kosong*). Another strategy was to install an 'ad interim' or caretaker (*pejabat sementara*) nominated by the village council with the approval of the district head (*camat*) as a temporary replacement when the term ended and no one wanted to run for election. In cases where the village had not decided to organise a village head election and no caretaker was suggested, the sub-district office would appoint someone as the interim head. The latter could be a military or police officer or a government employee from the sub-district office. Since they usually continued their current duties and did not live in the villages where they were appointed, these caretakers were rarely present in the village. This made it difficult for the villagers to obtain their services. However, there were only a few cases in the last decade of the New Order when a caretaker was sent from the sub-district office, because more than one third of the villages had an interim head who was nominated by the village council.

During the New Order, one of the functions of the village head was to ensure that Golkar won the village vote. One popular and successful way to do this was to promise villagers that streams of development projects would come to their village or to threaten that a Golkar loss would mean the end of 'progress'. The development of roads, schools, and health clinics was achieved by rotating the distribution of development funds and projects to each village in the sub-district. The village head would then rotate the funds and projects to each hamlet in the village. It was the promise of 'bringing progress' that villagers used to evaluate the village head's achievements, which would then determine the village head's success or failure in the next election. Since funds and projects needed to be rotated among all of the villages in the sub-district, a village that received funds then had to wait for the next cycle. The longer the 'waiting period', the smaller the chance of the village head winning in the next election. Success in bringing 'progress' to the village would prolong the village head's term of office and a fresh election might not even be needed.

The primary and most steady source of village development projects was the small annual village development fund (*bangdes* or *dana pembangunan desa*). The most commonly used way to use the fund was to build *gorong-gorong* (small bridges) and to gravel the village's unpaved roads each year. The fund was used only to buy the materials because the labour obtained through *gotong royong* or *kerja bhakti* (community works) was unpaid community work for all of the men in the village or hamlet.

Until recently, the village head received neither salary nor office space. The only legal sources of income for a village head were a small portion of funds collected from land tax (*pajak bumi dan bangunan*) and fees for services needed by the villagers. The amount from both sources was extremely small. In general, villagers accept that village officials take a portion of development funds and projects, but still refuse to accept the absence of village development projects. This creates a requirement for the village heads to accumulate wealth from state resources through the continued influx of development projects to the village.

One could safely say that what the village communities in the region would like to have is a village head who can fulfil the villagers' aspirations by bringing progress to the village. This is a formidable task. To ensure the flow of state resources into the village, the village head needs to get closer to higher levels of the state apparatus. During the New Order, this would be managed through the Golkar network and would involve petty corruption at various levels of administration, hence more cash in the pocket of the village head. If the village head went 'too far' with this petty corruption, however, the village community would react by setting up opposition in the village, developing factions, and spreading gossip to prevent the corrupt village head from winning in the next election. But without some involvement in petty corruption, it would be hard to bring development funds and projects to the village. No one would be able or willing to personally bear the transaction costs. A few village heads in the region were somehow able to maintain a balanced position. They managed to bring regular development funds and projects to the village, but were not overly corrupt, thus allowing them to maintain village community support (*dukungan masyarakat*). These village leaders managed to prolong their terms of office.

Efforts to keep the office within the family line by passing the office to children and/or to close kin have resulted in more failures than successes. In a few villages, the communities have nominated one of the children of a former village head to run in the next village head election. However, the nomination is usually based more on the nominated person's active involvement in village and community affairs, such as sports, religious feasts and village projects and/ or administration. In other words, it is the quality of the nominee that matters more than kinship *per se*. The village communities would be supportive of the nomination of anyone with such qualities, and village community support is incredibly important in village head elections. During the New Order, a connection to Golkar was much more important than community support, but more recently, community support has been the determining factor. Even during the New Order, community support could not be totally ignored. To avoid a win by an empty box in the village head election, community support was obtained by selecting a candidate who had the ability to use his relationships with higher government officials, via Golkar, to bring development to the village.

In West Lampung, the uniform name *desa* for administrative village, which had previously been the official designation throughout the nation, was changed to *pekon*. The head of the sub-village or *dusun*, formerly known as *kepala dusun* or *kepala suku*, then came to be known as *pemangku*. The village head formerly known as *kepala desa* — but informally called *lurah* — was renamed as *pertain* in 1999–2000 in line with the new national trend toward regional autonomy, which gave more authority to the district level. All of the new terms were said to be the original *adat* (customary) terms used by the native Lampung communities in West Lampung prior to Indonesian independence in 1945. This is when the former village councils came to be known as *lembaga himpun pekon* (village representative councils) and *lembaga pemberdayaan masyarakat pekon* (village councils for community resilience). Another important change was that village officials such as the village heads, village secretaries, village council leaders, and heads of hamlets were given a monthly allowance by the district government. The annual village development fund, increased to Rp 5 million from Rp 3 million, did not need to be used only for physical infrastructure such as *gorong-gorong* (small bridges) and roads, but could also be used for the village administration's operational costs. Another change was that the village head's term of office was reduced from eight to four years.

Previously identified as part of the New Order, these village leaders now act more as if they are part of the West Lampung district administration. One example is that there is reluctance among villager leaders to show clear loyalties to a particular political party. With the new disconnection of village administration from the political parties as well as the provision of monthly allowances from the government, the official village leaders' attachment to the district administration was strengthened. The village leaders began to act as if they were low-level parts of the government apparatus and now paid more attention to district policies and affairs.

In 2000, the head of the sub-district of Sumber Jaya launched a new policy which stated that 2002 would be the end of caretaker office terms in all of the villages in the sub-district. He also announced that the sub-district office would send one of its staff to be the village caretaker, and that no more village-nominated caretakers would be approved. Villages that still had village-appointed caretakers had to hold new village head elections.

Case Studies of Village Politics

I shall now examine the dynamics of village politics as illustrated by actual village head elections and leadership. In these examples, aliases are used for both village names and individuals.

Elections in 'Sukakarya' Village

Sukakarya is one of the villages created by the early BRN transmigrants. The last elected village head, Sarman, ended his term in the mid-1990s. Since then, the village has had an interim or caretaker — a position occupied in the first year by the former village secretary (*carik*), Amin. Another caretaker, Otong, was appointed for the next few years until a village election was held in late 2002. Both Amin's and Otong's appointments were based on nominations by the *musyawarah desa* (village assembly), with the approval of the sub-district head. Amin's nomination was based chiefly on his experience and knowledge of village administration, since he had previously served as village secretary. For Otong, it was his activity in the New Order and Golkar youth organisations at the sub-district level that led to his nomination. Otong's appointment was made possible because of his father's intensive lobbying within the village and at the sub-district office.

Otong's father Darsi was an elected village head from 1964 to 1983, while Sarman, his successor, won the village head election against an empty box. Among the early BRN transmigrants, not many had a high school education, and Darsi was among the few that did. His active involvement in village administration and community projects amazed the elders who then supported him to become the village head. It was during his term that most 'progress' (like school and road construction) was brought to the village, which enabled Darsi to enjoy a very long term in office. When he resigned as village head, he managed to become a member of the district house of representatives — first in North Lampung and then in West Lampung when the latter separated from the former in the early 1990s. He represented Golkar until the national election in 1999 that brought down Golkar and lifted the PDIP and the middle axis parties. Darsi's prominent involvement in the military hunt against Warman (Darsi himself was explicitly targeted by Warman's *gerombolan*) helped him to establish contact with higher levels of government, the military, and Golkar. It is through these well-established contacts that he was able to take a Golkar seat in the district assembly.

However, later on Darsi's son Otong was sacked from his caretaker office by the village assembly which was comprised of the heads of more than ten hamlets and village councils and mainly consisted of village elders. The villagers were disappointed in Otong's performance because he spent most of his time taking care of his *agen bis* (bus ticketing business) for passengers to Java, but they also opposed his father's influence on village affairs. Darsi used his son's position to gather popular support for himself and Golkar in the 1999 national election. With the *reformasi* following the fall of Suharto's New Order and Golkar, Darsi suddenly lost his influential power in the village.

Following the sub-district policy to end caretaker appointments terms and require an election of a village head, an organising committee was set up in Sukakarya. Yet, surprisingly, no one officially registered with the committee as a candidate. The few who were interested or nominated by factions in the village were either unwilling or unable to pay the costs of an election. The village committee had calculated the total cost for the election, and the candidates were responsible for paying this cost which was comparable to the cost of a wedding reception. For its part, the sub-district office asked for nothing except the cost of the photocopying and/or printing of the required materials. No bribe (*pelicin*) whatsoever was needed for a candidate to obtain official approval of a nomination.

Still, until late 2002, no one was willing to register as a candidate. The village assembly then decided that the village would be responsible for the cost of the village head election. An equal amount of cash was collected from each of the households in the village, and each head of hamlet was made responsible for collecting the money. In return, instead of candidates proposing themselves, the hamlets would select their own candidate to be nominated for village head. From more than ten nominees, the village committee approved seven candidates, and the sub-district office approved three of these nominated candidates. The rest failed since they had only an elementary school education and, according to the district regulation, a minimum of junior high school completion is a requirment.

Amin, the former village secretary and current interim head, was among those who were rejected. This led to great disappointment in the village since Amin was the favourite candidate. Election day was postponed to allow the village committee to lobby the sub-district office to allow Amin to be a candidate. The head of the sub-district advised the committee to persuade Amin to sit for an examination *(ujian persamaan)* equivalent to that of junior high school. If he passed the exam he would get a junior high school diploma *(ijazah)* and be officially approved as one of the candidates. The village committee, village council, and head of the sub-district were supportive of this idea and willing to postpone the village head election. But, to everyone's surprise, Amin refused to take the test. His close friends said that he was frustrated *(patah hati)* and embarrassed to be openly seen as too ambitious. Most villagers agree that had Amin's candidacy been successful, he would definitely have won the election. In his decades-long tenure as village secretary, he was neither involved in serious corruption nor in other wrongdoings and had significant village community support *(dukungan masyarakat)*.

Since the money collected from all of the village households was insufficient to cover the costs of the election, the village council decided to pawn the village fishpond to the village saving and credit association. Sukakarya is among a few villages in the region in possession of such communal land. For several years to come, the village saving and credit association was expected to manage and reap the harvest of the fishpond, which was more than a hectare in size.

So the village head election went on with three candidates: Haryana; Odo; and Tatang. All of the candidates were young — in their 30s and 40s. Haryana was the head of a hamlet and the only one with a couple of years of university education. Otong was active in the village savings and credit association. Tatang was another son of Darsi, but had no leadership experience, and his candidacy was largely 'steered' by his father. While Haryana and Odo worked their own coffee gardens, Tatang sharecropped his coffee garden. Tatang's house — the same house used by his father during his term as the village head — was the busiest on the day before the election. Friends and relatives gathered, and cars and motorbikes came and went. The host generously served meals and drinks for the guests. It was as if the house was holding a party. Large photos of Tatang were stuck on the front of houses, car windscreens, and shops in the village. By contrast, at both Haryana's and Odo's houses, it was as if nothing special was happening, with only one or two kin and neighbours chatting.

With so many people crowded in his house, Tatang's confidence was high. On the morning of the election, half a dozen cars with Tatang's poster on the windscreen were picking up voters from all of the hamlets in the village, including the two hamlets of his rivals, and taking them to the village hall. His confidence was further boosted by the odds in the gambling market, which were two or three to one in favour of Tatang. It is important to note, however, that those who were involved in the betting largely came from neighbouring villages.

The voting was held from 10 a.m. to 4 p.m., and began with a speech and official opening by the head of the sub-district, followed by a detailed explanation of voting procedures by the committee. There were no campaign speeches from the candidates. The candidates sat side by side with their wives in the centre of the hall during the opening, and all of them went back home immediately after the voting commenced. The voting began with women and elders entering the hall and exchanging the vote letter *(surat suara)* distributed the day before the election for the voting form with photos of the three candidates on it. The voters entered one of more than a half dozen voting booths *(bilik suara)* to punch a hole in one of the candidates' photos on the paper and put the paper in a large locked box at the centre of the hall.

By 2 p.m., with no more voters entering the voting booths, the committee decided to start the counting. The ballot box was opened and the counting began. Each candidate appointed an official witness to ensure a fair count. The fairness of the counting was further enforced by the presence of sub-district officials, police and military officers, members of the village council, and anyone who wished to attend. A very small number of registered voters had abstained, and a few voting forms which were not properly punched were considered invalid. The result was contrary to the expectation of many, especially outsiders, as Harnaya convincingly won the election. In the next couple of days there were stories about those who had bet on Tatang losing their bank savings, coffee gardens or motorbikes. The few who had bet on Tatang losing gained a considerable amount of cash. Yanto, a local Chinese businessman, was said to have instantly won Rp 10 million.

Gossip that the villagers had deceived Tatang and his father Darsi soon spread. A few days before the election, key figures in the village had openly expressed their support of Tatang. Some of the villagers said that this was done to avoid humiliating Tatang's camp because that might have led to chaos or disturbances (*rusuh* or *ribut-ribut*) in the village. Intimidation and violence were things that Tatang's camp was said to be capable of if they were humiliated. However, by ensuring a fair (*jujur*) and clean (*bersih*) ballot there would be no reason for Tatang's camp not to accept the final result.

Odo's loss, on the other hand, was a surprise to no one, being largely due to the work of his own camp in persuading villagers not to vote for him. The night before the election and on the morning of election day, Odo's close kin informed the key figures in the village that Odo's candidature was a mistake. He was too young, economically unstable, and immature as far as leadership was concerned. Many felt it would be better to give Odo a chance to improve his family's economy and his leadership skills in order to be better prepared for the next village head election.

Elections in 'Ciptapura' Village

A couple of months prior to the village head election in Sukakarya, Ciptapura, a village about 30 km from the capital of Sumber Jaya, held its own village head election. Ciptapura was created in the 1960s by two groups of Sundanese who now lived in the two main, neighbouring hamlets in the village. Each group had a charismatic leader — Sujana in Sukawaras, and Takim in Ciptajaya. Both leaders were legendary for their leadership roles in organising the early migrants to transform the forested land into the present-day Ciptapura. Both Sujana and Takim were separately able to persuade the neighbouring Semendo village head to give part of their village territory to the new migrants. Sujana and Takim were active in providing assistance to the migrants who settled in the village. Initially, assistance was given by simply clearing the forest and distributing the cleared land to each individual who helped. Later, newcomers were given a host in the village who allowed them to work on a piece of land as *numpang* (using a plot of land for free), as a sharecropper, or as a hired labourer, enabling them to accumulate enough money to buy land of their own. The communities in the two new hamlets sought advice from either Sujana or Takim, who were both among the richest men in the village, and who each owned more than 10 hectares of coffee gardens and rice fields. Later, Sujana focused more on formal leadership of the village while Takim became an informal leader, regularly receiving fellow villagers who consulted him on supernatural things, such as asking for a good day to undertake different tasks or for help with healing severe sicknesses.

The settlement turned into an official village (*desa*) in the early 1980s. A village head election was held and Sujana won against an empty box. Sujana was also Golkar *komisaris* (commissioner) in the village, ensuring a majority vote for Golkar in the village until the 1999 national election when PDIP won. Sujana's term as village head continued until the early 1990s when he retired, in large part because of his wife's health problems. No village head election was held at the end of Sujana's first eight-year term, and the village council and the sub-district office simply agreed to let him continue serving in the position for four years after his term ended. The villagers regarded Sujana as an ideal village head. He acted as a father to the village by ensuring fair decisions on internal affairs. He was also said to have never touched the village funds, and he let the village councils and village assembly make decisions on village funds and projects. More than that, Sujana was recognised for his achievements in bringing government projects to the village. It was during his term that the village built a health clinic, a market, two elementary schools, and several bridges so that the village's unpaved road network could be reached by jeep. The village was also continuously selected as the site of demonstration plots (*demplot*) for various agricultural extension programs, and since the mid-1990s, the village had been one of the most productive and intensive coffee-growing villages in the region.

When Sujana retired in the early 1990s, the village council appointed Sudarto as the interim head and planned to hold a village head election a year or two later. In the 1980s, Sudarto had migrated to Ciptajaya from Central Lampung, where he had bought a plot of coffee garden which was sharecropped while he was involved in the lucrative business of cutting and selling timber from the remaining state forest nearby. Upon his arrival in the village, he was appointed by the village council as the assistant *babinsa* (village military officer), and his main responsibility was to keep the village market secure. He received a regular income from the village funds collected from the traders in the weekly village market. Sudarto was successful in doing his job, preventing the stealing and pick-pocketing that had frequently occurred in the village market prior to his appointment. His appointment as the market security guard, and later as the interim head, was largely due to Takim's endorsement. Sudarto had long been in close contact with Takim.

Sudarto somehow managed to prolong his term as caretaker for almost a decade. A few years after his appointment, when the sub-district office questioned his status as caretaker and suggested a village head election, he was able to persuade the village council and the heads of hamlets to sign a letter stating that the village had agreed to extend his term of office. With this letter, the sub-district accepted the extension of his term. Like Sujana, Sudarto was very active in bringing government projects to the village: roads were gravelled, bridges, schools and a market were rebuilt, and a land certification project was also brought to the village. Sudarto's achievements and leadership were well recognised, but when it came to the issue of morality, the villagers expressed nothing but disappointment. Sudarto kept all of the village funds and left almost no room for the village council to have a say in village projects. It was also noted that he did shameful things, such as selling the gardens in the state forest zones whose settler owners had been evicted during military operations during the 1990s, stealing the villagers' money to pay the cost of land certification, and continuing his illegal timber business. The list grew to include other forms of wrongdoing, from drinking, gambling and 'playing with women' (*main perempuan*) to asking for cigarettes or drinks from shops without paying. It was only the last of these things that Sudarto was reported to have done within his own village. The other forms of wrongdoing were said to have been committed elsewhere, making them difficult to verify. The only proof was his frequent absence. A story about Sudarto's brother being caught in an act of burglary and later burning someone to death near the town of Metro in Central Lampung was used by the villagers to suggest the possibility of Sudarto's involvement in criminal networks elsewhere in the region.

In addition to the sub-district policy of having an elected village head in all of the villages, the village head election in 2002 was also the result of conflict between Sudarto and Takim, which led to the end of Sudarto's long term support from the most influential informal leader in the village. One of Takim's sons was involved in a fight with a young man from a neighbouring village. Normally, in cases of youth fighting with no weapons involved, both parties would enter discussions to reach 'peace' (*damai*), and the injured party would receive an apology and compensation in cash equal to hospital costs. The peace agreement would indicate that the case was considered an instance of juvenile delinquency that had been taken care of by the community, rather than as a criminal act to be taken to court by the police. In Takim's son's case, his enemy's family demanded compensation amounting to more than Rp 1 million — well beyond the actual medical costs required to treat the injury. Sudarto, in his capacity as head of village, did nothing to persuade both parties to discuss a peace settlement, but instead reinforced the demand for compensation and obliged Takim's family to pay it. Many believe that, had the compensation been paid, Sudarto would have taken a portion of the payment for himself because he had done this before to others in the village. Due to his strong informal leadership, Takim himself was finally able to settle the dispute in a peaceful manner, but by then he had become so angry with Sudarto that he promised to topple him from the office of village head. Takim's statement was embraced with much delight by most of the Ciptapura villagers.

A village committee for the village head election was soon set up. Juhana chaired the committee in his capacity as the head of the village council, but there was another problem. Apart from Sudarto, no one was willing to become a candidate. Takim soon asked Ujang, one of his sons, to contest the election. Less than 30 years in age, Ujang was studying at a private university in Bandar Lampung and, as a result, was frequently absent from the village. A couple of months prior to the election, Ujang married a Semendonese girl from a neighbouring village. Since there was no news prior to the marriage and no wedding party, it was said that the marriage was for the purpose of the candidature because according to regulations, a village head must be married. Takim's and Ujang's next steps were to then approach key figures in the village to gain community support. There was no problem with this as many key figures in the village were more than willing to support Ujang.

It is interesting to note that both Sudarto and Takim actually nominated Hardi to become the next village head. Had Hardi agreed to run, both Sudarto and Ujang would have withdrawn their candidacies in order to ensure Hardi's election. Hardi was in his forties and had a good leadership record. He was the head of the hamlet of Sukawaras, and was an active and influential young leader of the village council during Sujana's term as village head. He was economically established, with more than 3 hectares of productive coffee gardens and a couple of plots of rice fields, and had managed to send his two sons to Yogyakarta and Bandung — two cities in Java that were known for providing a good higher education. In the early 1990s, Hardi and his wife Minah went to the state palace in Jakarta to receive a national award from President Suharto. They were treated as pioneers in the national family planning program because they only had two children. During Juhana's term as village head, his wife's problems with literacy and poor health prevented her from performing tasks as head of the local family welfare education program. Hardi's wife Minah acted as leader of the program in the village and actively represented the village at higher levels of government. Thus, Hardi and Minah were seen by the villagers as an ideal couple to hold political office. To persuade Hardi to accept the nomination, the village council was willing to issue a decree that the village would be responsible for all of the costs of the village head election. However, both Hardi and Minah refused the nomination.

According to his friends, Hardi himself was quite willing and ready to accept the nomination, but not his wife Minah and their two sons. Minah felt from experience that the tasks associated with being the wife of the village head would be unbearably exhausting. Another problem was that the position of village head was not a particularly lucrative one. Although all village heads in West Lampung received a monthly allowance, the amount (Rp 250,000) was relatively small — equal to the cost of merely 100 kg of milled rice. The cost of living for an established family was about two to three times higher, according to some of the village heads in the region. Even though it was acceptable for the village head to use annual village funds for his personal needs, it would lead to gossip. For this reason, Hardi's youngest son strongly opposed the idea of his father becoming a village head. According to him, if his father became the village head, any goods (household goods, vehicles, and clothes) that the family bought in the future would be gossiped about as if the family had used the village's money. In particular, he could not stand to hear any gossip that the cost of his own study was paid for with the village's money.

Thus, it was eventually Sudarto and Ujang who competed in the village head election. To cover the cost, Sudarto sold one of his cars and some of his coffee gardens, while Ujang sold his motorbike and pawned some of his father's coffee gardens and *sawah*. The village council decided that no village money would be used. On the morning of the village election day, a dozen jeeps, minibuses and lightweight trucks, with either Sudarto's or Ujang's posters stuck to them, were busy picking up voters from all of the hamlets to take to the village hall in Sukawaras. Sudarto was reported to have been very nervous and got drunk the night before election day. He rode his noisy (but fake) Harley Davidson motorbike from hamlet to hamlet, and said to anyone whom he met on the street that he would take note of those who did not vote for him, threatening that something bad could happen as a result. To cool Sudarto's temper, hundreds of villagers gathered in his house on the night before election day, cheering him up and indicating that they would vote for him. Takim's house, where Ujang also lived, was much less crowded. It was said later that the villagers deliberately kept themselves from openly showing their support of him.

The voting procedures were similar to those in Sukakarya. The candidates and their wives arrived at 9 a.m. and sat in the middle of the hall watching the final preparation. Sudarto looked calm sitting on a couch, while Ujang was clearly nervous and frequently went out of the hall. The voting began around 10 a.m. after the head of the sub-district's official opening speech. Again there were no speeches from the candidates. In speeches delivered at both Sukakarya and Sukawaras, the head of the sub-district stressed that, unlike before *reformasi*, the government now had no favoured candidate (*tidak ada lagi calon yang dijagokan pemerintah*). This time, villagers should follow their hearts (*mengikuti hati nurani*) and vote for the best candidate for their village. Ujang and his wife left for home right after the opening speech and prior to the commencement of voting. Sudarto's wife left early, but Sudarto sat relaxed on the couch, smoking, exchanging jokes with members of the committee, and teasing some of the voters. He left for home a couple of minutes before the lunch break.

Unlike the vote in Sukakarya, no bets were laid on which one of the candidates would win or lose. No one seemed to dare to bet for either Sudarto to win or

Ujang to lose. The betting was on whether Sudarto would be able to obtain 200 votes from the nearly 2,000 registered voters, and the odds were even. As in Sukakarya, but with far fewer participants, the betting in Ciptapura involved motorbikes and coffee gardens as well as cash.

At 3 p.m. the voting was completed and the counting began. When the votes were all counted, it was revealed that Sudarto got less than 200 votes. A party was held at Takim's house that night to celebrate Ujang's victory. A couple of weeks later, Sudarto was thought to have left the village to live elsewhere with his other wife. Takim had not only deposed Sudarto from the office of village head, he had also gotten rid of his village rival.

Village and Sub-District Politics

By the end of 2002, the leader and the secretary of the sub-district of Sumber Jaya were both promoted. It was these two men who had established the policy that no more villages in Sumber Jaya would have caretaker heads by 2003, and all would have village heads elected through democratic elections. The secretary of the sub-district was appointed head of a less developed neighbouring sub-district, whose head was promoted to the leadership of Sumber Jaya.

The previous head of the sub-district of Sumber Jaya was promoted to head an office at the West Lampung district level in the capital of Liwa. He was not really keen to take his promotion, as he much preferred to continue in his position as the head of the sub-district. The village heads in Sumber Jaya also preferred to keep him as head of the sub-district because, according to them, he was unlike other *camat* in treating the village heads more as colleagues (*kawan*) than inferiors (*bawahan*). Perhaps more importantly, he never unilaterally asked the village heads to deposit (*setor*) money at the sub-district office, nor did he take a cut (*potong*) worth a considerable portion of village projects as funds for his personal use.[2] In the official ceremony for the handing over of the *camat* office, all of the village heads of Sumber Jaya made a declaration to the district head (*bupati*) that they wanted the present *camat* to stay. Acknowledging their sentiment, the *bupati* persuaded the village heads to give the new *camat* a chance. If in the following couple of months, they still could not accept the new *camat*, then a replacement would be arranged. This was a warning to the new *camat* to treat the village heads as colleagues rather than inferiors.

With the replacement of two key figures in the sub-district office, the policy of having elected village heads in all of the villages in Sumber Jaya was weakened. Among the fourteen villages, two villages still had caretaker heads in early 2003,

---

2   Nonetheless, this by no means indicates that there was no petty corruption at all.

and in both cases they were former village secretaries. In the first village, Trijaya, the village committee scheduled a village head election for the end of 2002. The cost of the election was an issue because the candidates were expecting the village to bear the cost, just as they did in Sukakarya, while the village council wanted the candidates to be responsible for the cost, as they were in Ciptapura. At the beginning of 2003, the issue had still not been resolved. In the second village, Sindang Cahaya, the situation was different. No one was willing to nominate as a candidate and there was no village council initiative (as in Sukakarya) where each hamlet nominated a candidate and the village bore the cost. As a result, the villagers were quite happy to have an extension to the term of the current caretaker.

In the sub-district of Way Tenong, unlike Sumber Jaya, the extension of the caretaker terms of office faced no obstacle. As long as there was no one willing to nominate as a candidate for the village head election, a caretaker's term would be prolonged. Nevertheless, people were still attracted to the position of village head. Two village head elections were held in Way Tenong in 2002. In both villages the candidates were responsible for the cost of the election. In one village, Hendra, one of the candidates was the richest man in his village, an important coffee reseller in the region and the owner of a large shop. Many people were surprised by his decision to run, since the material gain from the office of village head would be nothing compared to his current business. Hendra eventually failed to win the election, to the surprise of no one. It is said that the position of village head is one that no one can get without money (*tidak bisa didapat tanpa uang*), but it is also something that money cannot buy (*tidak bisa dibeli dengan uang*). The winner of this particular election was an ordinary villager (*orang biasa*). What Hendra lacked and could not buy was the villagers' popular support (*dukungan masyarakat*).

The granting and withdrawing of popular support for village leaders has played a key role in village politics throughout the region. *Dukungan masyarakat* was given to individuals who were able to meet villagers' expectation of integrating the village into the state and bring 'development' to the village. Village leaders were expected to keep promoting resource flows to the village, otherwise the *dukungan masyarakat* would be withdrawn and given to someone else.

5. Resource Control, Conflict, and Collaboration

After discussing the ways in which villagers bring the state into the village in the last chapter, this chapter will explore the ways in which villagers in Sumber Jaya and Way Tenong have resisted or accommodated government attempts to exert greater control over people and resources. As Peluso, Vandergeest and Potter (1995) have noted, one of the social trends in the political economy and political ecology of forestry in colonial and post-colonial Southeast Asia has been the consolidation of state power over forest resources, labour and territory. Furthermore, government attempts at forest control have created conflict between state agencies and villagers over forest land.

Smallholders who farm the land inside state forest boundaries in the region have been seen by Indonesian forestry authorities as *perambah hutan* (forest squatters/encroachers/destroyers). Villagers knew that farming the land inside state forest boundaries was illegal, yet they continued to transform forests into agricultural fields. For the latecoming landless migrants and the children of early migrants who aspired to become smallholder farmers, squatting on forest land was a way to gain access to farmland through non-market relations. Local people's resistance to the efforts of forestry authorities to transform smallholder fields into plantation forests and, more recently, community involvement in 'forest management', are generally viewed as efforts to restrict resource extraction from this sort of peripheral area by central elites.

History of Conflict over Lampung's Forest Zones

Between 1922 and 1942, the Dutch administration gazetted forested land in lowland and highland Lampung as forest reserves. On paper, the Dutch administration classified nearly 1 million hectares of Lampung land as state forest zones (*boschwezen*). Local people were prohibited from farming and gathering forest products from the gazetted forest zones. Until the Japanese invasion in 1942, the Dutch were able to conduct field delineation and boundary pole demarcation of more than half of the gazetted forest zones. These delineated forest zones are still referred to by local people as BW land (*tanah* BW), after the 'BW' (*boschwezen*) signs marked on the boundary poles.

In the post-colonial period, the national forest authority redesignated these *boschwezen* as *kawasan hutan negara* (the Indonesian name for state forest zones). Although the designation of the new forest zones was simply a reclassification of the former BW land, the process took decades to complete. The process began in the 1970s and was completed in 1990 when the Minister of Forestry signed a decree on Lampung's Forest Land Use Plan (Tata Guna Hutan Kesepakatan). According to the *Basic Forestry Law* of 1967, the Minister of Forestry had the authority to designate the state forest zones based on provincial government planning. The provincial governor's first proposal for Lampung's forest zones was in 1977, and the second was in 1980. Assessing this process, it appears that the governor was the one who proposed the new forest land use, but in reality it was the provincial office of the Ministry of Forestry which made these proposals. Afterwards, field delineation was carried out, new boundary poles were installed, and the old boundaries were reconstructed.

Meanwhile, massive logging operations were being conducted throughout Lampung. Beginning in the 1960s, logging became the main forestry work in the province until the end of the 1980s, when there were no more forests suitable for commercial logging. Logging concession areas included portions of land that were later designated as Way Kambas and Bukit Barisan Selatan national parks in the mid-1980s. In addition to former BW lands, the forest authorities also granted forests on *adat* lands to logging companies. These were either designated as state forest zones or given to estate plantation companies after they had been logged.

When the Lampung forest land use plan was proposed in the 1980s, a considerable   portion of the proposed state forest zones was no longer forested because of (legal and illegal) logging, conversion to village settlements, and the expansion of smallholding fields following the flow of migrants to the province. These facts were ignored. The justification for including these non-forested lands in state forest zones was the *Basic Forestry Law* of 1967, which stipulated that 30 per cent of the country's land mass must be zoned in this way. In the case of Lampung, this meant that 1.2 million hectares of the province's territory was officially allocated to state forest zones. These zones were then divided between: conservation forest (*kawasan konservasi*), designed to preserve the flora and fauna in its natural habitat; protection forest (*hutan lindung*), with watershed conservation as its primary function; and production forest (*hutan produksi*) for timber production. In the conservation forest zones (Bukit Barisan Selatan and Way Kambas national parks), more patrols have limited further encroachments but have not prevented illegal hunting or poaching and the expansion of smallholder fields.

In the 1980s, following the designation of state forestry zones, and with little forest left to log, the eviction of forest squatters and a program of reforestation became the main forms of protection. From the early 1980s to the mid-1990s, thousands of families were evicted from protection forest zones in various upper watershed regions in the province. These regions included: Gunung Balak in the east; Gunung Betung, Pulau Panggung and Wonosobo in the south; and Sumber Jaya and neighbouring sub-districts in the north. Evictions were accomplished through a series of military operations. Between 1979 and 1996, 65,000 families (over a quarter of million people) were resettled to several sites in the northern lowland of the province (for example, Pakuan Ratu, Tulang Bawang, and Mesuji) through local transmigration programs (*translok, transmigrasi lokal*).

It is public knowledge that those who joined the transmigration programs were only a fraction of those who actually settled and farmed in the state forest zones. Those who farmed in the state forest zones but did not live there were excluded from the local transmigration program. From the 1970s to the mid-1990s, official reports indicated that 180,272 hectares of protection forest had been reforested. But evidence in the field indicates otherwise, as most reforestation projects failed to transform 'degraded' forests into plantation forests.

Plantation forestry was conducted both in protection forest zones and production forest zones. Various government forestry units were made responsible for reforestation in protection forest zones, while in the production forest zones the state-owned company (PT Inhutani) and some private companies were involved in the industrial forest plantation scheme known as Hutan Tanaman Industri (HTI). From the mid-1980s to the mid-1990s, it was reported that PT Inhutani and other companies, which controlled 239,000 hectares of production forest between them, planted 54,907 hectares of fast-growing trees and rubber, but their success was limited.

Not only did the government fail to clear the state forest zones of settlers and reforest the 'degraded' land, but the eviction of the forest settlers and attempts at reforestation resulted in prolonged conflicts between smallholders, the forest authority, and the HTI companies. Meanwhile, the conversion of forests to smallholder fields continued, as did illegal logging of the remaining forests. In the mid-1990s, at least 41.4 per cent of some 316,570 hectares of conservation forest was no longer forested and had 5,676 households living within its various boundaries. Additionally, 83.5 per cent of some 318,513 hectares of protection forest contained 36,349 households, and 81.5 per cent of 401,910 hectares of production forest contained 54,000 households.

Map 5-1: Forest Land Use Plan or Tata Guna Hutan Kesepakatan (TGHK) of Lampung Province, 1990.

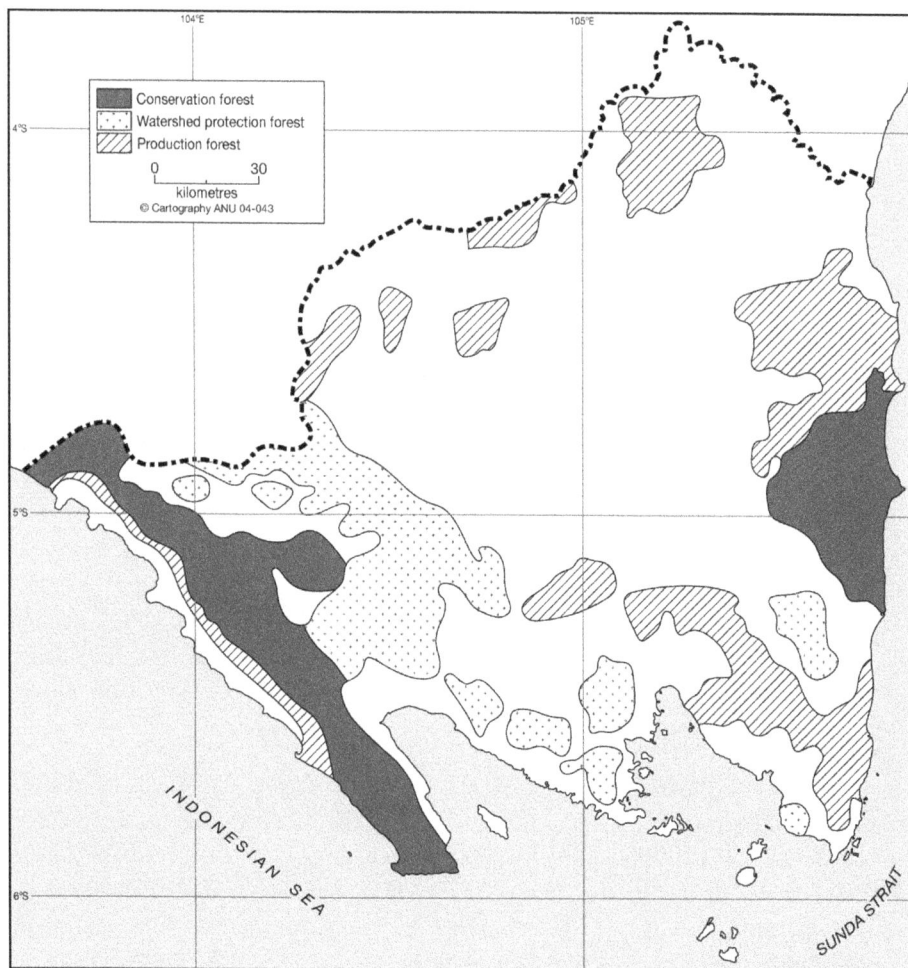

Land and Forests in Sumber Jaya and Way Tenong

In the Sumber Jaya and Way Tenong region, the state forest zones included all of the former BW lands, which already comprised a large portion of the region, but did not include the relatively flat land along the banks of the Way Besai River encircling Bukit Rigis Mountain. Located on the banks of the Way Besai, most villages in the region have state forest zone boundaries as their village borders. The former BW land to the east and north of the region was classified as protection forest. The hilly and mountain land to the west and north of the region was gazetted as part of Bukit Barisan Selatan National Park.

The opening of the BW forests began a few years after independence, and continued before, during, and after the creation of the province's forest land use plan. As early as 1946, on the border of what were then the sub-districts of Way Tenong and Bukit Kemuning, the forests were cleared for upland rice swidden, housing, and later coffee gardens, by a small group of Ogan and Semendo people from the neighbouring regions. Their hamlet, named Bedeng Kerbau, supplied rice for Indonesian soldiers who used Bukit Kemuning as their post during the revolutionary war in 1946–47. In 1965–66, the forestry service gave permission for 489 farmers to use 1,294 hectares of the BW land for housing compounds and farming, with each person receiving between 0.3 and 17 hectares. Finally, in 1969, the provincial governor officially recognised this territory as the administrative village of Dwikora.

From the early 1950s, more BW forests were transformed into settlements and farms for the transmigration of veterans from Java. As previously noted, the transmigration program was organised by a central government unit called Biro Rekonstruksi Nasional (BRN). The first BRN transmigration village of Sukapura possessed 224 hectares of BW land while another village, Tribudi Sukur, had 127 hectares. Elders in Simpang Sari remember that in the late 1960s there were already warnings from village officials to the incoming transmigrants to stop further encroachment onto BW land.

In the following decades, more and more people migrated to the region and as a result, more BW land was cleared and transformed into hamlets and smallholder fields. As more people led to more 'development' in the region, the creation of more administrative villages, the construction of roads, schools, and health clinics, and improved coffee prices and trade, even more people were attracted to the area. Yet enforcement of the forest zone boundaries was not an issue until the 1980s. Elders in Gunung Terang and Muara Jaya still remember that during the Dutch period, forestry officers regularly patrolled the BW boundary and advised village heads to deter their fellow villagers from clearing and farming land within the BW boundary. However, after independence the village head had no authority to prohibit outsiders from clearing BW land because the BW land was not part of the village territory and the incoming migrants were not 'citizens' of any village. Instead of asking the migrants to stop clearing the land, many village heads profited from these situations by charging the migrants fees, whether as a land tax or by granting farming permits (*izin garap*). The new migrants then became the 'citizens' of that village.

Illegal logging was another important factor that led to the deforestation of the region. In most villages the elites were — and a few still are — involved in this lucrative business. In the village, their responsibility was to organise felling, cutting, and local transport, and they were provided with protection by local policemen, military personnel, and/or forest rangers. This allowed them to avoid forest ranger patrols and raids, as well as police and forestry checkpoints along the highway when transporting the timber out of the region. It is a widely held view that timber is the main source of additional cash for local state agencies and their apparatus in those regions of Indonesia where alternative sources of additional cash, such as large industries or plantations, are absent. Without the backing of police, military or forest rangers, any local villager taking lumber, even from a naturally felled tree in the forest, will become the target of a local forest ranger's raid. The timber will be seized and the possessor will be sent to jail and released only after a sum of 'peace money' (*uang damai*) is paid.

Since the late 1970s, the region has been constantly targeted as a site for the sporadic implementation of forestry policies. Until recently, the main policy has been to evict smallholder farmers from state forest zones and to reforest their farmland, and one effective way of doing this was to accuse smallholder farmers of 'damaging the environment'.

The hilly region of Sumber Jaya and Way Tenong is the source of water for big river systems that feed large dams and irrigation schemes in the lowlands. This region forms the catchment for the Way Besai River, and is part of the larger Tulang Bawang watershed. The neighbouring regions of Tanjung Raja and Pulau Panggung are the upper parts of the Way Rarem, Way Seputih, and Way Sekampung watersheds. While the Way Besai hydropower scheme is an important source of electricity for the province, the Way Rarem dam supplies water for irrigation networks in the district of North Lampung. The Way Seputih irrigates rice bowls in the central Lampung lowlands, while the Way Sekampung feeds the Batu Tegi hydropower scheme in Pulau Panggung and irrigation networks in the southern lowlands of the province. For the forestry officers, the removal of natural cover and the planting of smallholder coffee on sloping land causes erosion, damaging the quality (siltation) and quantity (reduction) of water flowing downstream. Thus, smallholders have been accused of environmental destruction (*merusak lingkungan*) and were said to deserve harsh measures.

Reforestation projects in this region began in the late 1970s. They began with the planting of a few hundred hectares of pine and sungkai trees (*Peronema canescens*). A few stands of these trees can still be seen between Dwikora and Sukapura. During the 1980s and early 1990s, rosewood (*Dalbergia* sp.), calliandra (*Calliandra calothyrsus*), and mahogany (*Swietenia* sp.) were used in reforestation projects. Prior to the 1980s, projects were undertaken exclusively by the forestry service. After that time, the project would involve other parties, including private companies, PT Inhutani, and the military, which had its own reforestation project known as ABRI Manunggal Reboisasi (AMR). Previously concentrated only between the villages of Sukapura and Dwikora, reforestation projects spread throughout the region from the 1980s to the early 1990s, and most villages in the region received one.

In 1995–97, reforestation projects concentrated on the area around Dwikora and a few villages in the eastern part of Sumber Jaya, such as Sukapura, Simpang Sari and Tribudi Sukur, close to the site of the Way Besai dam. *Gmelina arborea* was the main type of tree being planted, along with a few other tree species. Private companies and the army were no longer involved in these projects, leaving their implementation to PT Inhutani and the forestry service. The reforestation projects often involved local men and women as paid workers, as well as labourers recruited from outside of the region.

Areas of 'bush' (*belukar*) and coffee gardens were the main targets for reforestation. It is reported that, between 1978 and 1985, reforestation projects planted trees in 20,000 hectares of forest zones in the region, and between 1995 and 1998, over 8,000 hectares were planted. After planting the trees on a particular site for a year or two, the project would move on to other sites and the newly planted trees were left without care. In areas of *belukar*, the trees soon died. In the coffee gardens, most trees were uprooted, but a few were kept alive alongside the coffee stands.

The eviction of farmers living inside of the state forest zones started in the early 1980s. Of the families whose coffee farms were demolished and planted with pine trees in Dwikora, a few hundred were involuntarily resettled in a new transmigration site in Mesuji, near Menggala in Northern Lampung. Hundreds of other families from the BRN transmigration villages of Purajaya, Purawiwitan, and Pura Mekar were among the more than 8,000 people targeted by a military operation to evict small farmers from state forest in Sumber Jaya and Pulau Panggung in 1990–91. They were also forced to resettle in Mesuji. In July 1994, in a joint operation of the forestry service and the police force, houses and coffee gardens in 86 hamlets in Purajaya, Purawiwitan, and Muara Jaya were destroyed. Some of the 1,200 affected families were resettled in Mesuji, while the rest moved elsewhere.

In other areas, to avoid the demolition of their houses and gardens by these military operations, the farmers dismantled their own hamlets themselves. Those who moved outside of the region simply abandoned their gardens, while those who lived nearby continued to maintain and harvest their coffee gardens. Living in a nearby village territory outside of the state forest boundaries, while continuing to care for and harvest the coffee gardens inside the state forest zones, became a common response to such military operations. Such a practice was called *kucing-kucingan* (hide-and-seek), and because it involved the pruning, pollarding, or felling of the reforestation trees, it had to be done carefully to avoid the risk of being caught in the act by patrolling officers who were becoming increasingly ruthless. During the harvest season, the patrolling forestry, military, and/or police personnel began to confiscate part of the harvest. Later, this became a regular practice, and every harvest season the farmers were asked to set aside a portion of their harvest for collection by the patrolling personnel who threatened to destroy their gardens if they were not compensated.

On the southern slopes of Bukit Rigis there were six hamlets, all located within the state forest zone. The population of three of these hamlets was registered in the transmigration village of Fajar Bulan, while the other three hamlets officially belonged to another transmigration village called Puralaksana. In the 1970s, they were populated by migrants, mainly from Java, and by the end of the 1980s, there were over 500 families within the six hamlets. Soon they were preparing to create a separate village administration. Sinar Harapan, literally meaning 'the light of hope', was chosen as the name of the village. Members of this community envisioned having a small market, an elementary school, and some mosques, just as in neighbouring administrative villages, but the plan never materialised. In the early 1990s, forestry and military personnel informed the villagers of an upcoming military operation to evict those living and farming in the state forest. Not wanting their houses and gardens destroyed, they vacated the village. Most of them moved to neighbouring villages. Their coffee gardens were soon planted with rosewood and mahogany. Only on regularly maintained coffee gardens have some of these reforestation trees survived; on the abandoned gardens and bushland they died.

The most recent government attempt to evict small farmers and turn coffee gardens and bushland within state forest zones into plantation forests took place in 1995–97. It began with what villagers in the region remember as 'the elephant operation' (*operasi gajah*) at the beginning of 1995. Unlike earlier military operations, this one involved a troop of elephants. The villages of Dwikora, in Bukit Kemuning sub-district, and Sukapura, Simpang Sari and Tribudi Sukur, in Sumber Jaya sub-district, were selected for the evictions. The start of the operation was aired nationally on the government television station (TVRI) and covered by local and national newspapers.

In a couple of months, the elephant operation managed to demolish hundreds of huts and houses, and thousands of hectares of coffee gardens. Unlike in previous operations, this time the villagers were more open in expressing their disagreement. Hundreds of Dwikora villagers organised a demonstration in the capital of the province. Delegates from this village also managed to engage in a series of dialogues with high-level provincial government officers and politicians. Petitions were signed and sent to key institutions in Jakarta, such as the Ministry of Forestry, the Human Rights Commission, and the House of Representatives. As a result, it was decided that houses located within 300 metres of a stretch of the main road more than a kilometre long would not be destroyed in the operation, but the villagers were still expected to dismantle their houses on their own, and demands for the cancellation or delay of the demolition of coffee gardens were rejected. In 1996, through a decree by the provincial governor, the administrative village of Dwikora was declared to no longer exist. By the end of 1996, a smaller troop of military, police, and forestry personnel was again set up to destroy the gardens, but this time hundreds of men wielding machetes rushed out and were ready to attack them. To avoid bloodshed, the government forces cancelled the demolition of houses and coffee gardens. The villagers only allowed them to demolish government facilities such as the village hall, water tank, and elementary school.

Resettlement and the further demolition of smallholder coffee gardens, as well as the planting of reforestation trees, still followed on from this exercise. In 1996, nearly 300 families from Dwikora and other villages in Sumber Jaya moved to Mesuji under a local transmigration program. PT Inhutani and various forestry units organised the reforestation project, and Dwikora was chosen as their base camp. By the beginning of 1998, it was reported that the project had planted at least 6,000 hectares around Dwikora.

Although the re-opening of previously destroyed and reforested coffee gardens had been occurring for some time, a massive re-opening began in mid-1998. In the early years of the monetary crisis, the price of export crops such as coffee and pepper increased sharply following the decline of the rupiah vis-à-vis the US dollar. Coffee prices rose fivefold from Rp 3,000 to nearly Rp 15,000 per kilogram. Dried bushland and dying reforestation trees were burned, making felling and clearing of bush and trees less arduous. The overthrow of Suharto and his New Order regime in May 1998 — marking the beginning of *reformasi* — was interpreted as the abrogation of the New Order's repressive forest policies, which allowed the *reformasi* to justify land reclamation and reappropriation. There was no more fear of forestry and military personnel. Apart from people reclaiming fields, a series of protests and demonstrations were staged throughout the province, resulting in some changes to forest policy. The news spread among villagers that there would be no more evictions and crop destruction, and that farming in state forest zones was no longer prohibited. The PDIP campaign for the 1999 general election centred around this theme, and the party's election win added further justification to the reclamation and reopening of state forest zones.

When talking about their interactions with the forest authority, villagers in the region often speak of a series of periods: *buka kawasan* (opening of [state forest] zones); *tutup kawasan* (closing of [state forest] zones); and *bebas kawasan* (free [to occupy state forest] zones). The first refers to the period prior to the enforcement of repressive forest policies; the second to the closing down of state forest zones from the late 1980s to the mid-1990s; and the last to the post-*reformasi* period. Some villagers said that if the coffee price not declined in 1999, the few patches of natural forest that still remain would be gone, completely transformed into coffee gardens. Certainly, throughout the region, most reforested lands were cleared and returned to coffee gardens.

Traditional Social or Community Forestry

The early model for reforestation projects in the 1980s can be seen as a form of social forestry. 'Forest farmers' were treated as free (unpaid) labourers in the establishment of plantation forests, and often the farmers worked as labourers on reforestation projects to plant rosewood in their own gardens. The farmers were allowed to keep maintaining and harvesting their coffee, but were strongly advised to look after, and not to fell, the reforestation trees, nor to abandon their gardens when the rosewood trees outgrew the coffee trees. Although some farmers followed this advice, most did not. They uprooted or felled the rosewood or kept only a few of them. Coffee gardens with a few rosewood trees can still be encountered in the region. However, most coffee gardens that were overgrown by rosewood and abandoned were soon taken over and transformed back into coffee gardens by other farmers.

During the late 1980s and 1990s, reforestation projects used harsher measures. Coffee trees were chopped down and the land was replanted with calliandra, gmelina, and other trees. Until 1997–98, thousands of hectares of calliandra bush covered state forest zones throughout the region, and gmelina trees were common in the eastern part of Sumber Jaya. By 2000, except for a few calliandra groves planted on bushland that was not suitable for coffee cultivation, most of this reforestation cover had been cleared.

Outside of the state forest zones, the forestry offices implemented 'people's forestry' (*hutan rakyat*) programs, which provided training, materials, and financial incentives for farmers' groups. The program included 're-greening' (*penghijauan*), in which fruit and fast-growing timber tree seedlings were distributed for free. Financial incentives for introducing farming techniques for soil conservation (terraces, ridges, pits), and the construction of small dams on creeks to reduce eroded soils flowing into river systems, were also included in the program.

In the mid 1990s, a different kind of 'social forestry' approach was put in place. This approach, which was very limited in scope, was concentrated in a few villages in the eastern part of Sumber Jaya, with the village of Simpang Sari being the main site. Farming blocks in state forest zones were grouped together, and villagers were employed as paid labourers to plant reforestation trees in their coffee gardens. Apart from exotic timber trees, a small number of non-timber trees were also planted. These non-timber trees, including petai (*Parkia speciosa*), aren (*Arenga pinnata*), jengkol (*Archidendron pauciflorum*), damar (*Shorea javanica*), and durian (*Durio zibethinus*) are officially known as multi-purpose tree species. The project itself was officially named *hutan kemasyarakatan* (community forestry). Forestry officials hoped that in the long run, the farmers would care for the reforestation trees, would be able to benefit from 'minor' forest products, and would give up cultivating coffee. In 2002, the farmers still cared for the young 'multi-purpose' trees, but the timber trees were uprooted, felled, or pruned regularly to prevent them from shading the coffee trees.

A few villages in the region were able to protect patches of natural cover adjacent to the village settlements. In the transmigration villages of Simpang Sari, Tribudi Sukur, and Cipta Waras in Sumber Jaya, as well as the Semendo village of Sukaraja in Way Tenong, a few hundred hectares of forest groves were prevented from being cleared and converted into farms. The forest grove in Sukaraja is an exception because it is located outside of the state forest zone boundary, unlike many other groves that are located within the boundary. This village forest is known as Kalpataru forest, after the national environmental award given to the Sukaraja community in 1987. Securing the water supply for rice fields and domestic use is said to be the villagers' top commitment. In all villages, the elders continue to remind the other villagers to protect the groves. Farmers — either from within or outside of the village — need land for farms. Aside from illegal logging by village elites backed by the military, police, and/or forestry officers, this need for farmland presents the greatest challenge for the villagers in protecting their forest. It is illegal logging operations and expansion of smallholder farms that have caused the failure of village forest protection in some transmigrant and Semendo villages in the region.

After the *Reformasi*

As far as relations between smallholders and forestry authorities are concerned, 'agro-forestation' and the protection of the remaining forest by local communities become a major policy theme in the region after the *reformasi*. Community forestry or *hutan kemasyarakatan* (HKm) was adopted as a program or policy that aimed to resolve the prolonged conflict over forest and land resources. The new policy marked the beginning of collaboration between forestry officers and village communities. However, the development of such collaboration has been problematic because the villagers understood the new policy to mean that there would be no more evictions and or destruction of their farms, while forestry officers saw it as a different strategy to gain greater control, not only over the resources but also over the people. Conflict over such divergent views is shown in the politics of resource control in the implementation of the community forestry program.

Plate 5-1: Members of a community forestry (HKm) group in Rigis Atas.

Source: Courtesy of the author.

At the provincial level, besides community forestry, *reformasi* in the forestry sector was also marked by a minor change in the forest land use plan, and the introduction of a regulation to impose a levy on all non-timber products (*iuran hasil hutan bukan kayu*) from all state forest zones in the province. The new land use plan (from 2000) excluded 145,000 hectares of production forest — mostly in the plains and lowlands of the province — that had long been converted to established village settlements, smallholder upland fields, wet rice fields, or brackish shrimp ponds on the coast. The new levy was designed to extract revenues from timber plantation companies which planted crops other than timber, as well as smallholders farming in state forest zones. For the smallholders, the exaction of the levy was linked to the granting of community forestry permits (*izin* HKm).

Under the community forestry scheme, smallholder farmers were required to form a farmers' group, or preferably a cooperative. The farmers' community group (*kelompok*) or cooperative was responsible for submitting a 'management plan' for a particular block of state forest managed by its members. The planting of trees — with a caution that coffee is not considered a tree — and protection of the remaining natural cover (if there was any) were the main components of the plan. The official contract of usufruct right was given to the group by the head of the district (*bupati*) and approval of the plan granted rights to the area for five years. After five years an evaluation was to be conducted. It was said that the results of the evaluation were used as the basis for granting more permanent permits which were valid for 25 years.

By the end of 2002, five community groups had been granted temporary permits. The farmers' group in Tribudi Sukur — which consisted of 15 smaller groups with 248 members managing 360 hectares of land — received substantial assistance from forestry office staff in 2001. For three other groups, assistance was also provided by field staff of WATALA and ICRAF. They assisted the groups in mapping and inventory, formulation of a management plan, and the granting of the temporary permits. Two of the farmers' groups were from two hamlets in Simpang Sari. One from Abung Marga Laksana consisted of four smaller groups with 73 members managing over 260 hectares of land, half of which was over-logged forest. The other one was from Gunung Sari, with 145 members managing 259 hectares land, including 90 hectares of over-logged forest. The third group was from Rigis Atas, a hamlet in Gunung Terang village, with three smaller groups managing 203 hectares of land, more than half of which had natural forest cover. The area of land per member for these three groups, and perhaps for other groups as well, was similar to the pattern of land control on *marga* (non-state forest) land which was between 0.25 and 4 hectares, with 1 hectare being the average. The last farmers' group to receive a temporary permit in 2002 was from the village of Tambak Jaya. Unlike the other groups, this farmers' group did not receive much help from external agencies.

The groups with temporary permits were responsible for the protection of the remaining forest from illegal logging. In addition, each village with a HKm group received a small monthly allowance from the district government for the appointment of persons — nominated by the group members and village leaders — as civilian forest rangers (*petugas keamanan* [or *pam*] *swakarsa*). This gave the authority in these villages to both the HKm group and the appointed *pam swakarsa* (civilian forest rangers) to stop illegal logging and the clearing of the remaining forest in the villages. After some initial raids, tree felling in forests near villages with an HKm group ceased. As some of the villagers involved in the illegal business said, 'we can no longer cut trees from the forest in some villages. It is now forbidden, not by officials (*petugas*) but by the community (*masyarakat*).' But forest protection by the village community created yet another problem. Timber now had to be imported from other areas, which led to an increase in the cost of home construction.

There were similarities among the various groups that were granted temporary permits. All of the groups were located on sites that were frequently targeted for evictions and crop destruction. With official permission, the villagers now had a more secure right to farm in state forest zones. As they often put it, 'we are safe (*aman*) now. We will no longer be the target of eviction and crop demolition.' Official permission was a strong motivator to join the HKm scheme. Among villagers themselves, there was sometimes conflict over 'ownership' of gardens on BW land where there were competing claims over a particular piece of land. Being registered officially with *izin* HKm ownership secured the land against any claim by fellow farmers.

Another similarity the groups shared was the large number of members of each group who lived in the same hamlet. As neighbours and friends and sometimes relatives, it was easier for them to form a group and to reach agreement on various issues. Strong leadership was another key issue as all of the groups that were granted permits had energetic, smart, and articulate leaders. The groups were therefore not only successful in reaching group consensus, but also in getting much-needed assistance from external organisations.

Being granted only temporary permits meant that the groups still had to obtain more permanent permits. This was a complicated issue since it was not clear how permanent permits were granted, and villagers had heard that there was reluctance in the forest authority to continue the implementation of the HKm scheme. There were reports that in many parts of the province, farmers were clearing more forests in anticipation of HKm permits because they had misinterpreted the policy as legal permission to convert more forest into farms. Villagers' resistance to official efforts to collect the levy on non-timber forest products imposed on smallholders farming state forest lands was another reason for a moratorium on the HKm policy.

Another problem which was more technical, but equally complicated, was the issue of planting trees. Villagers often had questions regarding how many trees needed to be planted, what species, and who should supply the seedlings. Smallholders who wanted to transform their coffee gardens into tree-based gardens were happy to plant as many trees as possible, but many were keen to keep coffee or other export crops, so they were more inclined to minimise the number of tall trees. Others wanted to plant more annual crops, such as vegetables, in their gardens, while trees that produced 'minor' forest products, such as fruits, sugar palm and resin, were strongly recommended for planting on HKm plots. Good quality seedlings were imported from outside of the region and were expensive, which was a major constraint. Some farmers also reported that their previous experience with planting commercial fruit trees (such as durian and longan) had shown that the harvest was too small and too irregular. Some crops like guava, jackfruit, and avocado fetched a low price. According to those in the village, it was the fast-growing timber trees that grew well and for which there was high market demand, but the planting of such trees was discouraged because, under the HKm scheme, forest farmers were obliged to plant trees but were prohibited from cutting them down, let alone selling the timber.

Farmers from a few other villages also formed groups in order to obtain temporary permits, but they were less successful due to a lack of skilful and trusted leaders and limited group cohesion. There were instances where the leaders of these HKm groups indicated their intention to secure personal gain from the group, which made other members reluctant to support them. In other cases, groups faced difficulties in reaching agreement simply because each faction within the group insisted on their own needs. In the worst cases, a group meeting was difficult to organise and so it was impossible to make a collective decision.

Given the large number of smallholders farming state forest zones in the region, the number of villagers engaged in the HKm scheme was relatively small. Many villagers made statements to the effect that 'the majority of people here in the region are forest settlers (*perambah hutan*) and most of the state forest zones in the region have been cleared and farmed.' This may be an exaggeration, but a large proportion of smallholders farming state forest lands in the region were not bothering to get official permits. For forestry officers, villagers who refused to join HKm and/or pay the levy 'lacked awareness' (*belum sadar*) of environmental conservation, 'needed education' (*perlu penyuluhan*), and were 'blind to the law' (*buta hukum*). The villagers, on the other hand, made equally valid points. The destruction of coffee gardens and the uprooting of reforestation trees became so frequent that, as they put it:

> We now are getting used to it (*sudah biasa*). It is a matter of who gets exhausted (*capek*) and gives up (*menyerah*) first. If we give up first, then they can plant timber trees. If they give up first, we continue cultivating the land.

Many villagers saw the conflict as a conflict over access to wealth. Illegal logging, reforestation projects, and premium-class or fast-growing timber trees were lucrative sources of income for the state. As some villagers put it, 'if all state forest lands are to be managed by the community, then how can those officers feed themselves (*bagaimana petugas bisa makan*)?'

The village of Simpang Sari is an interesting case that can perhaps represent the population in the entire region with regard to HKm. There were villagers who had successfully secured temporary permits, villagers who had formed groups but were still struggling to acquire permits, and villagers who did not want to be bothered with official administrative processes such as HKm. Those with permits were struggling to obtain tree seedlings and protect the remaining forest, and were confused over the additional burden of paying the levy. Other groups agreed to pay the levy but proposed that, in return, they be granted permits but without the obligation to plant trees and/or protect the remaining forests. Given such confusion, as well as internal leadership and cohesion problems, other villagers dissolved their groups and abandoned the HKm scheme altogether, while some felt that there was no need to join the HKm scheme and pay the levy because they had been paying land tax to the village administration for years.

The state of collaboration in forest land and resource management between villagers and forestry authorities is problematic, both in scale and substance. The protection forest zone of Bukit Rigis, for example, has a total area of 8,289 hectares. Heavily forested until the 1970s, about 2,000 hectares (less than 20 per cent) of the upper slopes remained forested in 2002, while the rest was mostly transformed into smallholder coffee gardens. Four years after the implementation of the community forestry policy, only a few hundred hectares of Bukit Rigis protection forest had cultivation permits. The process of obtaining the permits required strong community cohesion, exceptional village leaders, and/or external assistance, which more often than not has been unavailable. Under the community forestry scheme, smallholder households were required to form groups or cooperatives and make collective land use decisions instead. But this was problematic because individual households organised their agricultural production independently. Household farming decisions were made in response to the availability of farming inputs, market signals, and natural potentials and limitations, while the community forestry scheme demanded a management plan similar to the scientific approach used in the development of large-scale plantation forests.

Many villagers and a few forestry and other government officers believed that the remaining forests would soon vanish unless the nearby village communities protected them, while efforts to convert existing smallholder fields into plantation forests were unlikely to be successful. But they also well knew that those in power were very unlikely to hand over control of land and forest resources to the local people.

6. Gunung Terang: Social Organisation of a Migrants' Village

In Indonesia, legally each person and each parcel of land has to be integrated within an administrative village. This requirement was imposed by colonial administrations in Java (Breman 1982; Tjondronegoro 1984) and further strengthened in the post-colonial era. This nationwide integration was achieved by the introduction of the *National Village Law* of 1979, which imposed the adoption of a Javanese style administrative village (*desa*) throughout the archipelago. Within this context, ordinary Indonesian villagers were seen as members of an administrative community.

After discussing the region of Sumber Jaya and Way Tenong in the previous chapters, this chapter and the next chapter will discuss a single village in the region. Gunung Terang is an administrative village created by Semendo migrants in colonial times and afterwards inhabited by migrants from Java. This chapter explores elements of village social organisation in relation to village formation, leadership, and community cohesion.

In Sumber Jaya and Way Tenong, the administrative village functioned primarily as a vehicle to attract state resources to the villages. This fits the conceptual framework that positions local social organisations as intermediaries in rural development (Esman and Uphoff 1984; Tjondronegoro 1984; Quarles van Ufford 1987; Warren 1993). Along this line, Antlov (1995) suggests that under the New Order, rural leaders in Java based their power on administrative authority as state clients and/or on their ability to meet villagers' aspirations.

On the issue of village cohesion, Tjondronegoro (1984) notes that many communal tasks carried out by the rural communities in Java took place at the sub-village/ hamlet/neighbourhood level. Carol Warren (1993) has also noted the flexibility of Balinese villagers in organising themselves, depending on the nature of the tasks to be completed. Tjondronegoro (1984) and Warren (1993) further note that villagers' communal tasks range from planned development, to religious matters, to the household economy. It is in these ways that the residents of Gunung Terang have socially organised their lives.

The Creation of an Administrative Village

The village of Gunung Terang took its name from the oldest hamlet in the village. In this hamlet, the Semendo population is dominant. Most houses and fields in Gunung Terang hamlet are *tunggu tubang* properties passed down from parents to the eldest daughter. All of the Semendonese villagers in the hamlet are the descendants of *puyang* Tendak, a founding ancestress, as well as in-marrying wo/men (*jeme masuk*, incoming persons). Four generations ago, *puyang* Tendak's parents brought her and her two brothers from Ulu Nasal in Bengkulu, first to Mutar Alam and then to the new hamlet (*susukan*) of Gedung Surian. The new hamlet soon developed into a populous settlement (*dusun*) under the administration of Mutar Alam village. In the early 1940s, *puyang* Tendak and her husband Kemuli took their children and grandchildren and left Gedung Surian to open a new settlement at the present location of Gunung Terang. Their kin soon followed. The decision to migrate from Gedung Surian was largely driven by the need to find more land for rice fields, since there were not enough fields and no more land could be transformed into rice fields in Gedung Surian. In Gunung Terang hamlet, the riverbanks of Way Besai were transformed into rice fields, and families from neighbouring Gedung Surian and Mutar Alam, and also from Ulu Nasal (Bengkulu) came to settle in Gunung Terang.

Plate 6-1: Semendonese houses in Gunung Terang.

Source: Courtesy of the author.

According to a few elders, *puyang* Tendak was not supposed to leave Gedung Surian. As the only daughter, she was the *tunggu tubang* and thus entitled to inherit her parents' house and rice field. However, her brothers' refusal to observe the *tunggu tubang* rule forced *puyang* Tendak to find new land elsewhere. Some Semendonese in Gunung Terang hamlet believe that this refusal to recognise the *tunggu tubang* rule led to the Gedung Surian population suffering from illnesses and harvest failures so often that eventually the area was abandoned. In the

1960s there were only five Semendo families left in Gedung Surian, and the rest of the population had moved to Gunung Terang and elsewhere to survive. Some people in Gunung Terang use Gedung Surian's misfortune as an example of the punishment that comes from not observing the *tunggu tubang* rule.

In the Islamic month of Muharam each year, the Semendo community in Gunung Terang hold a *sedekah pusaka* ceremony. In the ritual, descendants (*keturunan*) of *puyang* Tendak gather in the house where the dagger heirloom (*keris pusaka*) is kept. The heirloom has been passed down from *puyang* Tendak to her eldest daughter, then to her eldest daughter's eldest daughter, and so on. Now the *pusaka* is kept by her great-great-granddaughter. The *sedekah pusaka* ritual involves reciting Qur'an verses and the cleaning of the *pusaka* dagger. Each family that joins the *sedekah* brings meals to be shared and served. According to some of the elders, the main purpose of the *sedekah* is to remember their origin (*asal usul*) and to ask God for his blessing (*berkah*) and for the well-being (*selamat*) of the community.

Like the Semendo, the Gumai — another Pasemah speaking group in highland Palembang (Sakai 1999) — stress the importance of an ancestry that places a person and a region as points of reference. Throughout the year, the Gumai perform many types of *sedekah*. The ritual specialists possess spiritual power and are highly respected in the ritual realm. This differs from the Semendo in Gunung Terang hamlet where the *sedekah pusaka* has been held only once a year, and was previously held only occasionally in difficult years such as during the Warman insurgency, or when there were harvest failures due to severe drought, or when there were epidemics of life-threatening diseases. The post-*krismon* drop in coffee prices and production after 1999, as well as post-*reformasi* political turmoil, has encouraged Gunung Terang residents to perform *sedekah pusaka* every year. The man who has been in charge of cleaning the dagger (*keris*) is regarded as one who knows how to do the cleansing properly but, unlike the ritual specialist in Gumai, he does not possess spiritual power and is not highly respected. The gathering on *sedekah pusaka* is arguably a way for *puyang* Tendak's descendants to maintain their social ties. Since most of them live in Gunung Terang, the ritual serves to strengthen community ties among the Semendo who live in Gunung Terang hamlet.

Map 6-1: Gunung Terang village.

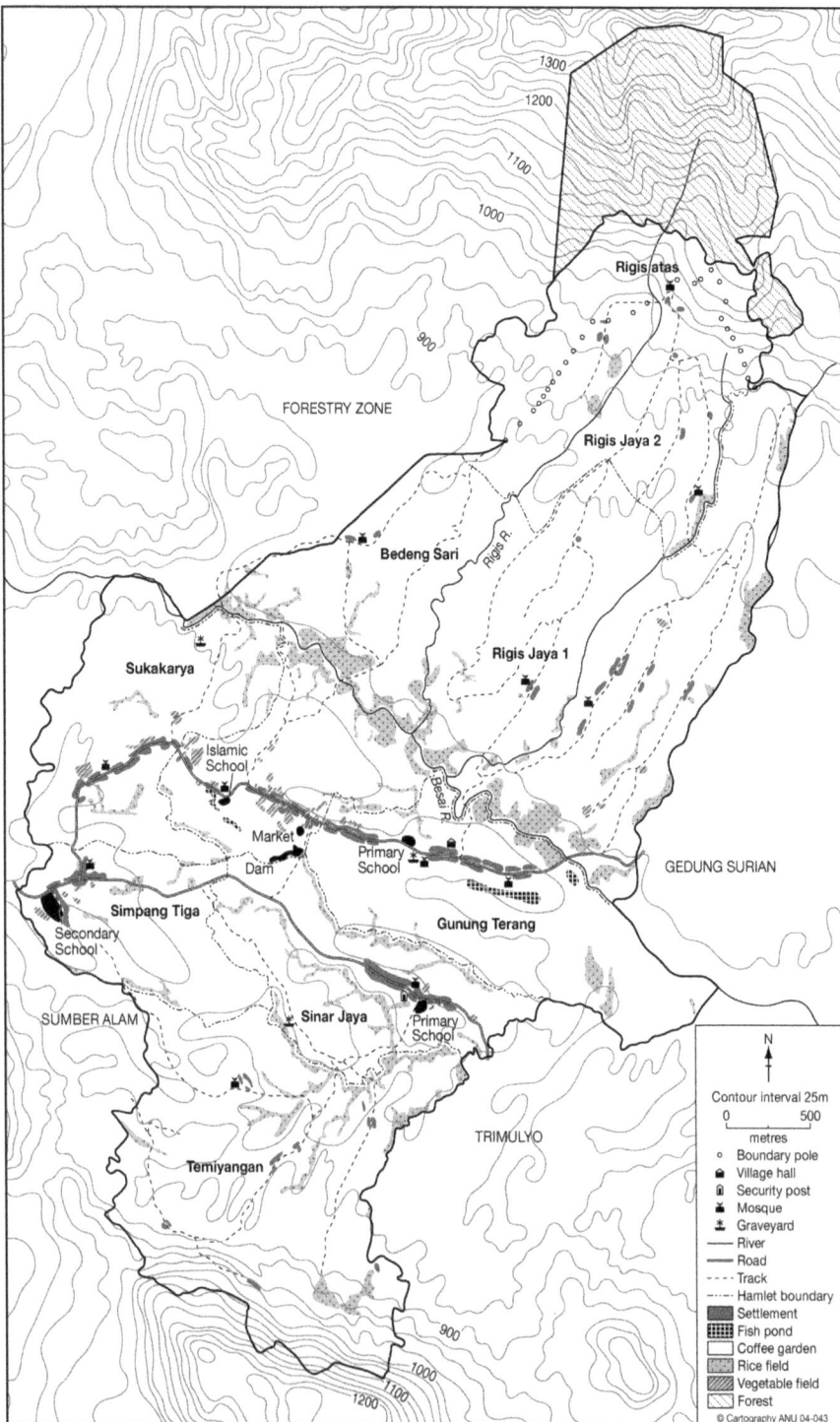

The map shows Gunung Terang village with the following labelled features:

- Rigis atas
- Rigis Jaya 2
- FORESTRY ZONE
- Bedeng Sari
- Rigis R.
- Rigis Jaya 1
- Sukakarya
- Islamic School
- Besai R.
- Market
- Dam
- Primary School
- GEDUNG SURIAN
- Simpang Tiga
- Secondary School
- Gunung Terang
- SUMBER ALAM
- Sinar Jaya
- Primary School
- Temiyangan
- TRIMULYO

Contour interval 25m
0 — 500 metres

Legend:
- Boundary pole
- Village hall
- Security post
- Mosque
- Graveyard
- River
- Road
- Track
- Hamlet boundary
- Settlement
- Fish pond
- Coffee garden
- Rice field
- Vegetable field
- Forest

© Cartography ANU 04-043

Other hamlets in Gunung Terang village (Table 6-1) were created between the 1960s and 1980s, mainly by migrants from Java. In Talang Jaya, the second oldest hamlet, there were initially less than half a dozen families from different parts of Java (Serang, Bantul, Nganjuk) who arrived in the late 1950s and early 1960s. They approached the Semendo in Gunung Terang and were allocated forest land, which they transformed into housing lots and coffee gardens, and turned local creeks into rice fields. More migrants from Java arrived and settled there, so by 2001, Talang Jaya had a fairly equal number of both Javanese and Sundanese inhabitants and both languages are spoken there. A few more Semendo people have also moved to the area.

Table 6-1: Population of hamlets in Gunung Terang village.

Hamlet	Population	Households
Gunung Terang	390	83
Bedeng Sari (incl. Talang Buluh Kapur)	875	176
Sinar Jaya (Talang Jaya)	239	71
Sukakarya (Petay Paya)	322	72
Simpang Tiga	183	48
Rigis Jaya I (Rigis Bawah)	429	123
Rigis Jaya II (Rigis Atas)	255	73
Temiangan and Talang Selingkut	238	60
TOTAL	2931	706

Source: Profile of Gunung Terang conducted by village administration, 2001.

The hamlet of Bedeng Sari has a rather different story. In the mid-1960s, a group of more than a dozen Javanese families arrived in Gunung Terang, including Pak Kono and his parents when he was 12 years old (in 1962). A native Lampung family sponsored their migration. Pak Kono's parents were assigned to take care of a citrus garden in Tegineneng, which was near Bandar Lampung. While caring for this garden, his parents also cultivated vegetables and raised goats. Despite good harvests, the situation in Tegineneng was difficult for the family because cash and goods were frequently stolen from their home. Five Javanese families in Tegineneng, plus another ten from different parts of Central Lampung, soon decided to migrate to Way Tenong. When they first arrived in Mutar Alam, they were advised to proceed to Gunung Terang, where the village head organised housing lots for these newcomers. He managed to persuade other Semendonese families to give the newly arrived Javanese land for housing lots (*kapling*). A bunkhouse (*bedeng*) made of bamboo walls and a grass (imperata) roof was built as a temporary communal house for them. From there, each family subsequently built huts in their allocated housing lots. The current hamlet took its name from the communal bunkhouse.

Plate 6-2: A house belonging to wealthy Javanese in Bedeng Sari.

Source: Courtesy of the author.

Labouring in the Semendonese coffee gardens and being paid in cash or in food (cassava, maize, rice, or bananas) was the main mode of survival for all of the newly arrived migrants. Access to land was also obtained by clearing the Semondenese forest or fallow plots in exchange for a portion of the newly cleared fields. Sharecropping was another way to accumulate enough money to buy a coffee garden. Of the dozen Javanese families who arrived from Hajimena/ Tegineneng in 1962, only five remained in Bedeng Sari in 2000, while the rest had moved elsewhere.

When the Javanese groups from Hajimena/Tegineneng arrived, there were already labourers and sharecroppers from Java living in scattered huts in the coffee gardens. The building of the bunkhouse (*bedeng*) and subsequent housing lots initiated the creation of more hamlets. More Semendonese lands along the path were sold at low prices to be transformed into housing lots for these sharecroppers and labourers, causing subsequent numbers of migrants to further extend the Bedeng Sari housing lots. Initially part of Talang Jaya administrative hamlet, Bedeng Sari separated and formed a single administrative hamlet. Later, Petai Paya and Simpang Tiga split from Bedeng Sari. In the late 1960s, Petai Paya was settled by two or three families who previously lived in BRN transmigration villages in Sumber Jaya before moving to Gunung Terang to work as labourers and sharecroppers in the Semendonese gardens. As in Bedeng Sari, the Javanese are dominant in Petay Paya, while Simpang Tiga is shared by the Javanese and Semendonese.

Until the 1980s, Rigis Bawah, Rigis Atas, and Talang Buluh Kapur were the locations of scattered gardens, patches of remaining logged-over forest, and fallow plots, many of which belonged to those living in Bedeng Sari and Gunung Terang. Javanese migrants, previously living elsewhere in Lampung, came to buy the land or to work as labourers or sharecroppers. Until recently, illegal logging has been an important economic activity in Rigis Bawah and Rigis Atas.

The hamlets of Rigis Atas and Temiangan have only recently been integrated into Gunung Terang's village administration. Rigis Atas residents previously lived in a hamlet within the state forest zone which has since been abandoned. Arriving in the late 1970s and early 1980s, the residents were part of the BRN transmigrant village of Puralaksana. When they were told to leave their homes and gardens in 1994–95 because they were going to be replaced with plantation forests, some of them moved down to the present Rigis Atas. After not receiving any 'attention' from the Puralaksana administration for years, the hamlet was integrated into the Gunung Terang administration in 2000. Like Rigis Bawah, Javanese are the dominant group in Rigis Atas, with Sundanese and Semendonese being the minority.

The hamlet of Temiangan used to be part of the neighbouring administrative village of Sumber Alam. Similar to Rigis Atas, the villagers in Temiangan also felt neglected by their village administration, leading the Javanese community in Temiangan to decide to become part of Gunung Terang village.

Plate 6-3: Houses in Rigis Atas.

Source: Courtesy of the author.

117

Recently there have been more attempts to mark the boundaries of the administrative hamlets in the village. A few Javanese houses on the edge of Bedeng Sari, for example, are administratively within the boundary of the Gunung Terang hamlet. However, they maintain day-to-day relations with their neighbours who are residents of Bedeng Sari. A suggestion to include these Javanese villagers within the Bedeng Sari boundary was rejected by the village council. Buluh Kapur, on the other hand, is administratively eligible to form an administrative hamlet separate from Bedeng Sari, yet residents there wished to remain part of Bedeng Sari.

A hamlet is socially and territorially divided into several neighbourhoods. Both hamlet and neighbourhood have a communal responsibility or role. Community works (*gotong royong*) for paths, roads, small bridges, running water tanks, and pipes are sometimes undertaken by all of the residents of the hamlet, but at other times are done only by the neighbourhood men. Most hamlets have a mosque (*masjid, mushalla*) that is constructed and communally maintained. The residents gather in the mosque for regular Qur'an reciting (*pengajian, yasinan*), Friday prayers, and to celebrate Islam's holy days. Some neighbourhoods have smaller prayer houses (*surau*).

Adults in Gunung Terang village are keen to be seen as devoted (*taat*) Muslims. Before sunset, men wear a sarong and cap (*peci*) and prepare for evening prayers. Most of them do the daily prayers in their homes, leaving the *surau* and mosques empty. Friday is the weekend in the village, and on this day villagers stay at home. The villagers come to the mosque for Friday prayers and a speech (*khutbah*) at mid-day. The *khutbah* consists of the imam reading a section from a book containing a collection of Friday speeches (*buku kumpulan khutbah Jum'at*). School-age children in Bedeng Sari and Petai Paya go to a small Islamic school (*pesantren*) to learn Al Qur'an reading, Arab script writing, and Islamic teachings (*ajaran*). In other hamlets this is done in the *surau* and the mosque in the afternoon or evening. Women form *pengajian* groups and meet once or twice a week in the *surau* or the mosque to recite Al Qur'an and hear Islamic teachings.

Community cohesion at the neighbourhood level is stronger than at the hamlet level. The Javanese in the village admit that, with regard to helping a member of the hamlet experiencing hard times (for example, death, illness, accidents, and personal conflicts with outsiders), cohesion (*kekompakan*) among the Semendo community in Gunung Terang is exceptionally strong. For religious feasts (*sedekah* and *ruwahan* among the Semendo, and *selametan* or *syukuran* among the Javanese and Sundanese), it is the neighbours' obligation to give *sumbangan* (a gift or donation) of raw food (for example, rice, sugar, chicken, or coconut) and snacks such as biscuits. Close neighbours also help in the preparation of the feast, and women who are close neighbours and kin usually help with the cooking of the meals. In the case of a death, the burial and the subsequent prayer rituals would be the neighbourhood's responsibility. It is quite common for a villager to have close neighbours (*tetangga dekat*) who are also good friends (*kawan baik, akrab*). Among the poor, the bond between close neighbours is particularly strong. Often their huts or wooden houses have been constructed communally, and they tend to organise a reciprocal labour exchange which limits their need to hire labour. Among themselves they often arrange zero-interest credit partnerships or form rotational savings groups (*arisan*).

Like villages elsewhere in the region and other parts of rural Indonesia, Gunung Terang has two patterns of housing — nucleated (or compacted) and dispersed (or scattered). In the hamlets of Gunung Terang, Talang Jaya, Bedeng Sari, Petai Paya, and Simpang Tiga, the houses are nucleated and aligned in a row along the village's main road. All of these hamlets, except for Talang Jaya, obtained electricity in the late 1990s. In Buluh Kapur, Temiangan, Rigis Bawah, and Rigis Atas, houses are dispersed along the unpaved roads and paths. In Rigis Atas and Rigis Bawah, there are a few small compact housing compounds with up to a dozen houses separated by coffee gardens.

Villagers refer to the construction of facilities and the provision of services by the government when discussing the progress (*kemajuan*) in their village. Suharto's New Order era of the 1980s is said to have been the turning point in village progress. As some villagers put it, 'before there was nothing (*tidak ada apa-apa*) in the village, everything was difficult (*payah*), and life was hard (*susah*).'

Clinic and health programs for women and children have been some of the most recognisable measures used to distinguish the difficult (*susah*) years before the 1980s. Sick, pregnant women and infants often died before arriving at the clinic in Fajar Bulan. They are now quickly taken care of by the nurse in the village or in the health clinic (*puskesmas*) in the neighbouring village of Sumber Alam. In serious and/or emergency cases, patients are cared for at the small hospital in Fajar Bulan. The well-being of women and infants has been further improved thanks to periodic *posyandu* (short for *pos pelayanan terpadu*, or integrated health service posts). These are clinics at which sub-district nurses and village family welfare functionaries provide information to women on infant health issues, vitamins, and immunisations. Many families in the main hamlets in Gunung Terang once received sacks of cement from the government to improve their housing.

Progress in the village during the New Order extended into the economic sectors. In the mid-1980s, hundreds of families received generous agricultural extension assistance, and extension officers regularly visited the village to advise the farmers on better farming techniques. Incentives in the form of cash, tree seedlings and livestock were provided to encourage the application of soil conservation measures on sloping land, and credit was provided in the form of chemical fertiliser and tools. The annual output from coffee gardens, rice fields and other farmland increased dramatically. The improvement in agricultural production was followed by a cheap land certification project that allowed villagers to use the certificate as collateral to obtain loans for various purposes from the Bank Rakyat Indonesia branch at Fajar Bulan.

In Gunung Terang village, facilities and infrastructure were equally distributed among the main hamlets of Gunung Terang, Bedeng Sari, Petai Paya and Simpang Tiga. The village now has two elementary schools, one in Talang Jaya and one between Bedeng Sari and Gunung Terang. The village's Islamic school (*pesantren, madrasah*) is located in Petai Paya, as is a health clinic run by a nurse. The junior high school (*sekolah menengah pertama*) is located in Simpang Tiga. The village hall (*balai desa*) is located in the hamlet of Gunung Terang, while the village weekly market is in Bedeng Sari.

In the 1980s, a road network was built to connect the villages on the southern part of Bukit Rigis with the West Sumatra Highway, which passes across the northern slope of Bukit Rigis. The village paths, previously constructed by village communities through *gotong royong*, were enlarged, gravelled, and asphalted, and the wooden bridges were replaced by sturdy concrete bridges. Before the construction of the road network, sacks of coffee beans had to be carried manually to coffee resellers in Fajar Bulan, which could take a whole day. To obtain household supplies, one had to walk to the weekly market in Srimenanti and later to Fajar Bulan. This meant walking to the market the night before market day and sleeping over in the market. Since the construction of the road, Fajar Bulan can be reached in less than an hour by motorbike, pickup, or minibus. The road was asphalted in the mid-1990s. A weekly market recently opened in the neighbouring village of Sumber Alam, where each Friday over a thousand people from neighbouring villages come to do their weekly shopping.

Before the construction of the current schools, children in Gunung Terang went to Mutar Alam. A few literate adults in the village voluntarily taught the children in the basement of a stilted house in the beginning, and then later in wooden classrooms that were built through *gotong royong*. The government finally developed this informal school into a formal elementary school in the 1960s. The second elementary school in Talang Jaya was built in a similar way. When the sub-district education office asked the villages to grant land for the construction of a secondary school in the early 1990s, the villages of Gunung Terang and Sumber Alam were quick to agree to grant land on the border between the two villages. Because of this, children only need to go to Fajar Bulan for high school education.

Progress has been a source of tension between the Semendonese and the Javanese in the area. On the surface, one might consider such tensions to be ethnic conflicts, but they are perhaps better seen simply as manifestations of the desire for progress.

As a Javanese man put it, 'the Semendonese here are difficult (*payah*), they don't want our village to flourish (*ramai*)'. Two cases of the reluctance of the Semendonese to release some of their land for the sake of village progress can be used as illustrations. Unlike some of the neighbouring villages, the main hamlets of Gunung Terang, Bedeng Sari, Petai Paya, Simpang Tiga, and Talang Jaya are separated by over a hundred metres of coffee gardens. This prevents the hamlets from being conjoined to form a larger village settlement. Some of the Javanese villagers suggest that this is largely due to the Semendonese reluctance to allow their gardens to be bought by fellow villagers and new migrants and transformed into house lots.

Another Javanese man criticised the Semendonese for not allowing their land to be taken to enlarge the current path from the asphalt road to the hamlet of Rigis Atas. Had the Semendonese agreed, it would have reduced the transportation cost and attracted more new migrants to the remote hamlet of Rigis Atas. This allegation is, of course, denied by the Semendonese. They say that their reluctance to sell the land between the hamlets is due to the fact that the village is now full of migrants from Java, so if they sell the remaining land it would be difficult to find another plot that they could buy as a replacement. The prohibition on selling *tunggu tubang* property is another constraint. The Semendo also maintain that the cancellation of the road to Rigis Atas had more to do with financial problems, and the technical difficulty of constructing a bridge crossing the Way Besai River, than with getting land to enlarge the existing path.

For their part, the Semendo complain that the Javanese are always trying to sideline them. In the eyes of some Semendonese, Javanese domination in the village will put the Semendonese in danger. All of the government projects go to the Javanese while the Semendonese are left behind (*ditinggalkan*). Such tension occurs especially between the dominant Javanese hamlet of Bedeng Sari and the old Semendo hamlet of Gunung Terang. The tension has led to talk of either Bedeng Sari or Gunung Terang splitting to create a separate administrative village.

In the 1960s, the administrative village of Gunung Terang included the present neighbouring villages of Tri Mulyo, Cipta Waras, Gedung Surian, and Semarang Jaya (Air Hitam). Tri Mulyo was created by a group of Javanese villagers whose leader, Sumardi, had lived in Talang Jaya for a couple of years before moving on and founding Air Dingin, the main hamlet in Tri Mulyo. Sumardi brought 17 families with him from Central Java in the 1960s. Pak Cik Nawi, the village head of Gunung Terang, gave them permission to clear the forest there, and Air Dingin hamlet was soon followed by others. These hamlets officially became the separate administrative village of Tri Mulyo in the mid-1980s. Pak Cik Nawi also gave permission to two groups of late BRN transmigrants from West Java to settle in the area. One group of a dozen families from Tasik Malaya, led by Pak Juhana, first came to Tribudisukur only to find that there was no more land available in this BRN transmigration village. This group then created the hamlet of Waras Sakti. The other group of about 40 families, mainly from Bogor, first came to the village of Puralaksana, but it had no more land available for them either. With Pak Cik Nawi's consent, this group cleared the forest and created the hamlet of Ciptalaga. Initially both hamlets belonged to BRN transmigration villages (Waras Sakti being part of Tribudi Sukur, and Ciptalaga part of Puralaksana), but in the mid-1980s, both hamlets and the neighbouring hamlets formed a separate administrative village called Cipta Waras. Pak Juhana was elected as the first village head.

The abandoned hamlet of Gedung Surian soon filled up with Javanese and Sundanese migrants. It also separated from Gunung Terang's administration in the 1980s, retaining its old Semendo hamlet name. A portion of Gunung Terang land was also given to hundreds of families from Semarang (in Central Java) who arrived in Mutar Alam in the late 1970s and early 1980s. They settled in Air Hitam and later created the village of Semarang Jaya.

Village Leadership

When one asked villagers in Gunung Terang about the people who could best tell the history of the village, they mostly pointed to three men: Pak Kasijo in Talang Jaya; Pak Timan in Petai Paya; and Pak Cik Nawi in Gunung Terang. Aged in their sixties and seventies at the time of my fieldwork, these three men were former village leaders and were now considered to be village elders (*sesepuh, tokoh*).

Pak Kasijo was well known for his prominent role in promoting children's education in the village. He arrived in Lampung in the mid-1950s from Bantul, near Jogjakarta. Initially he planned to join his relatives who had migrated to Wonosobo in southwestern Lampung, but upon arrival in Lampung he took up an offer to work as a foreman (*mandor*) in a rubber plantation and factory in Kotabumi that had formerly belonged to the Dutch, but had since been nationalised. After a couple of years, he decided to migrate to the newly opened transmigration area in Sumber Jaya and Way Tenong. With his friend Sumardi, who later led a group of Javanese villagers to open the land in Trimulyo, he finally settled in Gunung Terang.

Unlike most early migrants, Pak Kasijo was a high school graduate. He opened the first community school in Gunung Terang in the 1960s. At that time, children had to go to Mutar Alam to get an elementary education. Pak Kasijo's initiative was very much welcomed by the villagers. He and a few other villagers with junior high and high school education voluntarily acted as teachers. Initially the classes were held under the stilt house of Pak Cik Nawi, the village head. Later, through *gotong royong*, all of the villagers worked together to build a simple wooden house as the classroom. The community school later became the first formal elementary school in the village, located between the hamlets of Gunung Terang and Bedeng Sari. Pak Kasijo then continued his efforts to open a second elementary school in the village located near his house in the hamlet of Talang Jaya.

Pak Timan was another elder well known for his leadership in village affairs, especially agricultural extension. In 1962, as an orphan boy, he was taken by his uncle to Kota Gajah in central Lampung. When his uncle's family broke up (*berantakan*), he was taken in by an indigenous Lampung family to maintain their pepper garden. Given a hard time by the children of his Lampung foster father, young Timan joined a friend who left for Sumber Jaya in 1964. For a few years he lived in Simpang Sari, farmed a small plot of coffee garden, and married a Sundanese girl whose father was an early BRN transmigrant. Later, Pak Timan decided to move to Gunung Terang to join (and later replace) his brother, who was working to maintain one of Pak Cik Nawi's coffee gardens. Pak Timan eventually managed to establish his own coffee garden and build a decent house in Petai Paya hamlet.

Pak Timan and his wife were active in village affairs. His wife was active in assisting Pak Cik Nawi's wife in various family welfare programs, like the *posyandu* program. When a group of migrants from Hajimena/Tegineneng arrived in Gunung Terang, Pak Timan became involved in the construction of the communal bunkhouse (*bedeng*). He was active in creating the hamlet of Bedeng Sari, but Pak Timan's outstanding leadership was ultimately recognised for the work he did in organising villagers to receive government assistance in agriculture during the 1980s. He was chair of a village farmers' group (*kelompok tani*) with over two hundred members for more than a decade. With an elementary school education, his literacy was an important reason the villagers chose him for this position. Pak Timan coordinated the provision of credit for members of the *kelompok* to buy chemical fertilisers from the sub-district agricultural extension officers who also routinely gave advice on better planting materials and cultivation techniques. Pak Timan was often selected as the farmers's delegate in meetings in the capital of the sub-district, district, and province. In his capacity as chair of the *kelompok*, he initially organised the villagers in the land titling program. According to Pak Timan, two thirds of the village population now had their land titled.

The most prominent leader in the village of Gunung Terang was Pak Cik Nawi. Unlike Pak Kasijo and Pak Timan, who were now retired, Pak Cik Nawi was still active in village politics. Born in Gedung Surian, he was still a little boy when his grandparents and their offspring moved to create Gunung Terang hamlet. He learned about village administration mainly from two relatives — the then village heads of Gunung Terang and Mutar Alam. Pak Cik Nawi was first appointed as the interim village head in 1962. He won the village head election in 1965, lost it in 1972, but was reappointed as caretaker head two years later. He failed to win the village head election in 1979, but was again appointed as caretaker head in 1983. In the 1990 village head election, his candidature was rejected by the sub-district office, mainly because his level of education was lower than the minimum requirement (junior high school graduate). Since then, he had been a key figure in the village council.

Many of the early migrants from Java in the village likened Pak Cik Nawi to a 'parent'. He was remembered for his efforts from 1960 to 1970 to ensure that each migrant family had a house to live in and land to work with. Some early migrants from Java still remembered how, during difficult years, Pak Cik Nawi allowed them to take rice from his rice field and other food (like cassava, banana, and jackfruit) from his garden. He persuaded some Semendonese villagers to do the same, and actively persuaded the Semendonese to welcome the newly arrived migrants from Java.

Most of the migrants were initially landless, including those who worked in his gardens. Pak Cik Nawi earned respect for not treating them as inferiors nor taking much material advantage from his position as village leader. Pak Cik Nawi possessed an average amount of wealth, and was not among the handful of truly wealthy families in the village. None of his children went to university, and he could not even afford to send his two younger sons to high school.

Pak Cik Nawi's role in the village during the late 1990s was mainly that of an adviser to village officials on village affairs. He was formally the chair of the village council. Some villagers exaggerated this role by saying that 'without Pak Cik Nawi's approval (*restu*), village projects could not be implemented smoothly'. This did not mean that all of the village projects would be successful, even if they did have his approval, nor does it suggest that he had the power to impose his opinion on village decisions. Rather, it was recognition of his persuasive ability to encourage key actors in the village to come to a consensus. Development projects, such as the construction of schools, roads, bridges, and water networks, require villagers' participation for their successful implementation. Negotiations over such projects can easily be the source of tension between factions and/or sections in the village. Most often the tensions were between the Semendonese and the Javanese, and Pak Cik Nawi's advice was mainly directed at resolving these tensions.

Village head Bu Mas Muda, village secretary Mas Paryoto, and village council chair Pak Cik Nawi were three key figures in the village administration. Gunung Terang was the only village in the region led by a woman, Bu Mas Muda. Her good leadership was recognised not only in the region of Way Tenong and Sumber Jaya, but also in the whole district of West Lampung. The villagers were proud to have her as the village head. One of the hamlet heads in the village once proudly claimed that 'no village head in the region or elsewhere that I happen to know is better than our village head, Bu Mas Muda'.

Bu Mas Muda won the village head election in 1998. She was quite underestimated by her only rival, a man from Simpang Tiga. Apart from her ability to gain full support from the hamlet of Bedeng Sari where she lives, her success in the election was also due to her ability to win the votes of the Semendonese in the village, who lived primarily in the hamlet of Gunung Terang. She promised that the Semendonese would not be 'left behind' in the village development projects — an issue that especially worried the Semendonese if the village head was Javanese. She learned much about village administration and affairs, and also ways to bring development projects to the village, during her husband Pak Hasan's term as Golkar village *komisaris* and village head between 1990 and 1998.[1]

Recognition of Bu Mas Muda's leadership was largely due to her success in bringing development projects to the village. Between 1999 and 2001, there were several such projects as part of the package of loans and grants that the Indonesian government received from international development agencies to cope with the 1997–98 *krismon*. The path through Rigis Bawah was enlarged and gravelled to enable car transportation. The road from Simpang Tiga to Talang Jaya was also gravelled, while another local road construction project shortened the distance between the villages of Trumulyo and Fajar Bulan. A network of plastic pipes to supply running water was installed from the Bukit Rigis foothills to the hamlets of Gunung Terang and Bedeng Sari. For the villagers, especially the poorer ones, these works provided substantial wage earnings. For her success in bringing such development projects to the village, Bu Mas Muda was respected by the villagers.

Preventing tension among village sections was another item on her working agenda. By nominating Bedeng Sari and Gunung Terang as the intended

---

1   In the 1990 village head election, Pak Hasan won by only 30 votes against an empty box.

beneficiaries of the running water project, she prevented tension between the two hamlets. Gravel — the main material for road construction in the hamlet of Rigis Bawah — was supplied by those living in the hamlet of Rigis Atas. The latter received the payments while the former received the improved road, so both hamlets enjoyed the benefit of the project.

Plate 6-4: Residents of Rigis Bawah doing *gotong royong*.

Source: Courtesy of the author.

Although active in seeking government projects to be implemented in the village, Bu Mas Muda was quite careful not to put the village (and her leadership) in a potentially difficult situation. A small group of men, including some hamlet heads in the village, once intended to engage in a soft credit scheme provided by a government agency of the district. The credit would have been distributed among households in the village to buy compost, fertiliser, and pesticides for commercial vegetable farming. A well-written proposal was prepared and was ready to be submitted. Intense communications had been established with district officials who would provide the credit, so the chance of obtaining the credit was deemed to be high. Yet Bu Mas Muda gently refused to approve the initiative. She pointed out that the village had had bad experiences in handling a government credit scheme in the past. The village had received cheap credit under an IDT program, which was used to buy sheep and goats that were distributed selectively to poor households. The credit was designed to rotate among the poor. Soon all of the sheep and goats were reported to be sick or dead, and a number of village officials were taken to the district attorney's office and accused of corruption. Although no one was proven guilty, it was a great humiliation for the village and its officials.

But despite her refusal, Bu Mas Muda did not totally ignore the proposal. She supported the idea of forming a village vegetable farmers' group, and agreed to allocate village funds for village delegates to visit some 'advanced' (*maju*) vegetable farmers in neighbouring Sekincau. The newly formed group was to collect cash from each of its roughly 40 members, who would then decide how the money was used and monitor its use by the membership. Bu Mas Muda wanted to see if the farmers' group — of which she herself was a member — possessed the ability to handle its members' money before trying to engage in risky credit schemes provided either by the government or by private agencies.

According to many villagers, Mas Paryoto, the village secretary, had leadership skills in village affairs besides being able to process village administration paperwork. Before being appointed to this position by Bu Mas Muda, Mas Paryoto was appointed by the sub-district office as the village enumerator (or data collector) for the family planning and social welfare program (*petugas pencatat keluarga berencana desa*). Maintaining good communication among village officials and leaders was a task that Mas Paryoto managed quite well. He regularly visited formal and informal leaders in the village and kept them informed of village affairs. He maintained close contact with all of the hamlet heads in the village, either by visiting them, often with Bu Mas Muda, or inviting them to his or her home. In this way they were both were kept informed of things happening in all of the hamlets, while the hamlet heads were informed of government policies and programs related to village affairs.

Mas Paryoto was active in promoting commercial vegetable farming in the village. With his two neighbours, he started the commercial and highly intensive cultivation of vegetables. It started with capsicum chilli, and then other vegetables such as tomato, eggplants, and beans were also introduced. Villagers frequently gathered in his house to hear his technical advice on how to start commercial vegetable farming. He was also frequently invited to see fellow villagers' vegetable fields and give suggestions. He made contacts with traders or salesmen of agricultural inputs for commercial vegetable farming (seeds, fertilisers, chemicals, and so on), and a small group of villagers gathered frequently in his house to hear a salesman promote his products. Mas Paryoto kept persuading the salesmen to give free demonstration samples, allowing a variety of brands to be tested and compared. He was also one of the initiators of the newly formed village farmers' group whose purpose is to assist members with growing better commercial vegetables through the provision of agricultural inputs and better marketing of outputs.

With his skill and ability in village administration, maintaining good communication among village leaders, and promoting commercial vegetable farming, some villagers believed that Mas Paryoto was the most suitable candidate to be the next village head when Bu Mas Muda's term ended in 2006. But Mas Paryoto was reluctant to be seen as too ambitious. Secondly — and this was a far more serious concern — he felt that his family was not yet economically established (*cukup, mapan*). According to him, an economically established family is one of the prerequisites for an ideal village head. In 2002, Mas Paryoto was still struggling with his family's economy, and therefore decided not to run for village head. He did not own a coffee garden, but was a sharecropper who took care of less than 1 hectare of coffee garden belonging to another villager. He had just started farming commercial vegetables in his 0.25 hectare house garden. He was still not sure whether he could afford to send his two little daughters for higher education. Although his father was a large landowner in Rigis Bawah, with over 10 hectares of coffee gardens, and had been able to support Mas Paryoto's high school education in Java, the land was equally distributed among his children from two wives when he died. Mas Paryoto's share was then sold to buy a house lot and to build his present house.

Each 'administrative' hamlet in the village had a head, but unlike the village head, all of them were appointed by the hamlet residents by consensus (*musyawarah*) instead of being elected. Since 2001, village officials had been receiving a monthly allowance from the district government. Among the administrative tasks of the hamlet heads (*pemangku, kepala dusun, kepala suku*) are those of recording monthly and annual data on the demography of the hamlet and collecting the annual land tax (*pajak bumi dan bangunan*). They represent the hamlet at village meetings and are responsible for delivering messages about new government policies from the village administration to the hamlet community. Within the hamlet they are expected to maintain social harmony (*rukun, tentram, guyub*), which includes the task of settling disputes amongst neighbours, giving advice on official matters, organising the hamlet's religious rituals (such as *yasinan*, the celebration of Islamic holy days, and burials), and encouraging *gotong royong* activities to construct and maintain community facilities.

There were variations in the leadership role of the hamlet heads in the village. In the hamlets of Gunung Terang and Bedeng Sari, the role of the hamlet heads was rather limited and focused only on the collection of demographic data and land tax. These were the two hamlets in which Bu Mas Muda, Mas Paryoto, and Pak Cik Nawi resided, so the villagers in both hamlets heard about village affairs from them directly rather than through the hamlet heads. In 2001–02, the hamlet heads of Bedeng Sari, Gunung Terang, and Talang Jaya had little involvement in community work. In other hamlets, such as Rigis Bawah, Rigis Atas, and Temiangan, where the residents only rarely met higher ranking village officials, the leadership role of the hamlet heads was more prominent.

Like other village officials, hamlet heads were recognised for their efforts to integrate the community into the village administration and tap state resources. The name of Talang Jaya hamlet was taken from the name of its first head, Pak Jaya, who put a lot of work into creating the administrative hamlet and integrating it into the village of Gunung Terang in the early 1970s. Pak Maryono followed a similar strategy in the 1980s in what was now the hamlet of Rigis Bawah, formerly known as Talang Maryono (a name which was still used informally). Pak Maryono was a key figure in Golkar's success in the village during the general elections from the 1980s until the 1990s. As Golkar *komisaris*, he actively persuaded the Javanese villagers — now the majority people in the village — to vote for Golkar. Pak Simun, the then hamlet head of Temiangan, had more recently taken a similar role in separating the hamlet from the village of Sumber Alam and integrating it with Gunung Terang. Mas Kaulan, the then hamlet head of Rigis Bawah, gained his reputation thanks to the recent government road building project in his hamlet and a community water supply project. Conversely, Muayat Wagimin, the then hamlet head of Rigis Atas, organised his community to supply the gravel for a road building project in Rigis Bawah which provided much needed extra paid work. But his most prominent leadership role was in organising the hamlet's residents who farm the state forest zone to engage in a community forestry agreement (HKm).

The official village administrative structure had other posts as well. Under the village head, there were several heads of special affairs (*kaur*, short for *kepala urusan*). Under the hamlet head there were also several heads of neighbourhoods (*kepala rukun tetangga*).[2] The village council had about a dozen members and the village also had several civil security officers (*hansip*, short for *pertahanan sipil*). Aside from village officials responsible for witnessing and recording marriages, divorces, and reunions (*petugas pembantu pencatat nikah talak dan rujuk*) and neighbourhood heads in some hamlets, these were mostly nominal offices. The position of chair of the family welfare program was usually occupied by the village head's wife, but in Gunung Terang the position was given to another woman who had previously assisted Bu Mas Muda when she was chair during her husband's term as village head.

The religious leaders in the village, including the board of the mosque (*pengurus masjid*) and teachers and preachers in the village's small *pesantren*, concentrated on religious teaching and rituals. Each mosque in the villages had one or more

2 In Buluh Kapur, which is part of the administrative hamlet of Bedeng Sari, the neighbourhood heads were active in *gotong royong* and religious rituals.

imam or *kiyai* whose role in everyday affairs was minimal. Functionaries from national political parties in the village had no great role in everyday village affairs. There was no sign of political activity from the major parties' *komisaris* and cadres other than putting up signboards and attending party meetings in the sub-district or district.

The village had no official office. The village hall was only used during village meetings attended by higher level government officials. For internal village official meetings, Bu Mas Muda's house was used. The village paper work was done at the village secretary's house where Mas Paryoto had a study. All of the village officials in Gunung Terang were full-time farmers, and consequently part-time village officials. They worked in their gardens in the morning and afternoon every day. They came home for lunch and *dzuhur* (mid-day prayers) during the working week, and except for busy seasons in the farming calendar, they were at home the entire day on Fridays. Within this time-frame, any villagers wanting to see the village head and village secretary had to find them early in the morning, during the mid-day break, just before sunset, or else in the evening.

A Village Development Meeting

In June 2002, the village held a village development meeting (*musyawarah pembangunan pekon*). The meeting was supposed to be the venue for the village community to outline the village development plans they wanted the government to fund. The meeting was organised at the request of the sub-district office (*kecamatan*). About a week before the meeting, an official letter of invitation was sent to village officials, members of the village council, hamlet and neighbourhood heads. The letter was signed by Bu Mas Muda and distributed by Mas Paryoto himself. Mas Paryoto also visited all of the hamlet heads personally to promote the occasion and advise them to be prepared. In the following days, many hamlet heads and a few other village officials were seen visiting either Bu Mas Muda or Mas Paryoto to further discuss the preparations for the village meeting.

The meeting was held at the village hall. The delegates had already gathered at about 9 a.m. but had to wait for a couple of hours to start the meeting. This was a rare occasion for villagers from different parts of the village to meet and have a lively chat. Nearly 80 delegates attended the meeting. Adult males were dominant. There were numerous young men but less than a dozen women. The meeting began when the district officer arrived at nearly 12 o'clock. Mas Paryoto opened the meeting by greeting all of the delegates, explaining the purpose of the meeting, and outlining the agenda and timetable. The meeting had three main agenda items: explanation of the new government policy on village administration; selection of the chair and members of the village community development council (*lembaga pemberdayaan masyarakat pekon*); and a workshop on village development plans that concluded the meeting.

Bu Mas Muda delivered the opening remarks. She began her speech by stressing that each hamlet should propose development programs that were deemed to be urgent (*penting, mendesak*) and actually needed (*dibutuhkan*). She reminded all of the village hamlets to collect the targeted amount of land tax on schedule; distribute the government-subsidised rice (*beras miskin*) only to those who were eligible, such as poor families; and not to wait for an order (*perintah*) to undertake *gotong royong*. The second speech was by Pak Cik Nawi in his capacity as the chair of the village representative council. First, he advised delegates that, in selecting the village community development council members, they must choose among those who lived permanently (*menetap*) in the village. He continued his remarks by stating that the village conducted village development meetings like this every year, but that the results of those meetings were never followed up. The village development plans only ended up piled high (*menumpuk*) at the district government office. Yet because there was a formal request from the sub-district office, the village must again hold a village development meeting. Pak Cik Nawi concluded his remarks by repeating Bu Mas Muda's comments that the delegates should only propose programs that were urgently needed.

The district officer began his section of the meeting by explaining the new government policies in accordance with the newly enacted national, provincial, and district laws, regulations, and decrees on village administration and development. He continued by explaining that the village should now have two councils with complementary roles. The village representative council was responsible for formulation, ratification, and enactment of village decisions and regulations. The village community development council functioned as the working partner (*mitra kerja*) of the village official administration. Gunung Terang had already selected the members of the first body but still needed to select villagers for the second one, which has a chair, a deputy chair, a secretary, and a treasurer. There were to be eight sections in the village community development council: (1) religion and community harmony (*kerukunan warga*); (2) legal institutions (*kelembagaan hukum*) and laws and regulations (*perundang-undangan*); (3) youth, sport, art, and culture; (4) improvement of human resources, natural resources, and environment; (5) economic development; (6) family and women's empowerment; (7) media, communication, and information; and (8) customs and tradition (*adat isitadat*).

The meeting continued with the election of members for the village community development council. One of the criteria for candidates for chair of the council was that he or she must live on the main road so the chair would be accessible to the villagers when needed. Since the hamlets of Rigis Atas, Rigis Bawah, and Temiangan were not along the main road, they were not allowed to nominate a candidate. All of the delegates were given a small piece of paper to write down the name of the candidates. Nurdin, a Javanese, had the highest number of votes and became the chair, while Ka'i, a Semendose, came second and became his deputy. The secretary and treasurer were appointed by the village head, the chair of the village representative council, and the elected chair of village community development council. The positions in charge of the eight development council sections were filled with candidates from all of the hamlets.

The next agenda item was the discussion of the village development plans. The sub-district officer started by explaining that financial sources for village development programs and projects were available from three levels of government — central, provincial, and district. The funds would be divided and used for programs and projects by all *liding sector* (leading sectors) implementing agencies at the district level. Before beginning the discussion, the sub-district officer expressed his dissatisfaction with the DPRD members from the region. He said that although the Sumber Jaya and Way Tenong region had many DPRD members, none of them 'fought' for development in the region. Instead of backing up sub-district officers at the district level, they just did the '4Ds' (*datang, duduk, diam, duit* or 'came, sat, were quiet, and sought money'). As a result, he noted that there were piles of proposals for village development plans sitting on desks at the district government office — a reiteration of Pak Cik Nawi's remarks. Yet because the district government required the sub-district office to submit village development plans that were actually proposed by the village community, this kind of village meeting had to be held.

The sub-district officer continued by reading the list of 24 *liding sektor* with which the village proposed plans should be matched. The officer read out the name of each sector and the delegates then mentioned a plan that might suit that particular sector. Most often he proposed the plans himself, and they were then accepted by the crowd with a loud 'Agree!'. He also frequently rejected plans proposed by the delegates if he thought they were irrelevant. For example, when delegates proposed a project for running water, it was simply rejected because Gunung Terang had already received such a project. When the discussion came to the irrigation *sektor*, a delegate proposed an irrigation project. The officer then asked whether the village had an intact area of more than 50 hectares suitable for rice fields because the government would not fund any irrigation projects if the area suitable for rice was smaller than this. Since none of the delegates could answer his question, the proposal for an irrigation project was simply erased from the list. The discussion became a bit lively when it came to the agriculture and natural resource management sector. In response to the decline in coffee prices and recent forest clearing, an agricultural diversification program was proposed that included vegetable farming and tree and cash crop planting. The discussions focused on what crops would grow well in the area and have good market demand. Proposed projects for each *liding sektor* continued in this fashion until the workshop was completed less than two hours later.

The meeting concluded with the sub-district officer reminding the audience of the allocation of the annual village development fund (*dana pembangunan desa*). Since there had been so many allegations of corruption, the district government decided to stop providing this fund in 2000 or 2001, but the village heads had complained that they could not run the day-to-day village administration without the money. So the district government had decided that the annual village fund would again be provided, but that it must be used only for the operational costs of village administration. To use the annual village fund as a source of credit for income-generating activities or for physical infrastructure, as was done in the old days, was now totally prohibited.

The village development meeting finished at about 2 o'clock. Lunch was provided for all of the delegates. They all went home without knowing what would happen with the village's proposed development plans, but they all knew that at some time in the following year, a very similar village development planning ritual would be organised, and they would again submit to doing what they had just done that day.

Contingent Cohesion

As a corporate group, an administrative village is characterised by clear membership and territorial boundaries. The village consists of several hamlets, and each hamlet is made up of several neighbourhoods, each of which might consist of only half a dozen families or households. Official village affairs tend to be taken care of by the village and hamlet, while other community affairs (like religious or emergency matters) are organised by the hamlet and neighbourhood.

At the village level, community cohesion is seen within the context of formal state 'rituals'. The village development plan meeting, for example, is a venue that serves to symbolise the existence of a village community, and the participation of the village community in national rural development planning is then required. No one expected that the result of village planning would seriously be considered and followed up by the higher-level government decision makers. Nevertheless, the gathering itself strengthened the delegates' feeling that they were members of an entity and were discussing matters that would benefit all of its members.

The celebration of 17 August — the nation's independence day — is another occasion where the villagers' sense of community is accentuated. For Independence Day in 2002, the sub-district office asked all of the villages to maintain tidiness (*kerapihan*) and raise the national flag. Following this request, the national flag and colourful banners (*umbul-umbul*) were erected in front of houses along the main road, with bamboo fences all painted in white. Sporting matches and games (volleyball, checkers, and dominoes) were held for a week. The celebration concluded with *panjat pinang*, where boys competed to climb the greased trunk of a palm tree to collect small prizes hung on top of it. The games were entirely a village initiative, and the money for the celebration was collected from all of the households in the village. Here again, the gathering served as a venue for the villagers to meet and do things together. Besides strengthening the sense of community at the village level, these 'rituals' also deepened the villagers' feeling of being part of the larger Indonesian national community.

Maintaining the village graveyard is another activity in which the villagers act as a community. About a week before the fasting month of Ramadan, each hamlet sends about a dozen men to weed the village's main graveyard located in the hamlet of Gunung Terang. Although not all deceased villagers are buried here, men from every hamlet are involved. The village graveyard itself is not very large and it could easily be weeded by less than a dozen men in less than half a day. Yet more than three dozen men gathered to weed it in 2002, proving that it was not the weeding itself that was important, but the fact that villagers from all of the hamlets in the village took part.

In the old days, the village *gotong royong* was the primary means to get a village project done. Through years of village projects, paths were enlarged and wooden bridges were constructed so that motorbikes or four-wheel jeeps could pass by. The use of the village annual development fund to buy material for the small bridges (*gorong-gorong*) was almost a necessity. The construction of elementary schools and the village market were also done this way, and many villagers from all of the hamlets spent days and even weeks on the projects. Often, when additional money was needed to buy materials to complete a project, cash was collected from all of the households in the village.

While successful in building mosques in all of the big hamlets, the villagers' plan to have a *pesantren* in the village had yet to materialise. Through donations and *gotong royong*, the biggest mosque in the village was built at the border of Bedeng Sari and Petai Paya and a few classrooms were attached to the mosque. Following the drop in coffee prices, donations ceased to flow, and the plans to create more classrooms for the school and boarding houses (*pondok*) to house pupils from outside of the village did not eventuate.

The road networks of Bedeng Sari and Buluh Kapur are examples of unsuccessful village projects. Only motorbikes could navigate these roads. The problem was not in mobilising *gotong royong* to enlarge the path, but in getting the cash to purchase materials for some small bridges that needed to be constructed in order for the road to be passable by car or jeep. This upgrade would ultimately reduce the transportation cost for goods such as coffee beans, fertiliser, and building materials. Coffee gardens in Buluh Kapur and Rigis Atas do not belong exclusively to the residents of the two hamlets; many belong to villagers in the hamlets of Bedeng Sari and Gunung Terang. But despite the urgent need, there was no serious plan for road construction.

The construction of a weekly market was also an unsuccessful village project. The market, which operated on Wednesdays, failed to attract as many traders and buyers as the Thursday market at Ciptalaga and the Friday market at Sumber Alam. Even the Gunung Terang villagers themselves prefer to go to Sumber Alam for their weekly shopping.

Plate 6-5: A shop (*toko*) in Sumber Alam village market.

Source: Courtesy of the author.

The recent project for water supply is another example of failure. The project was heavily subsidised by the government. The government (in this case the Department of Public Works) provided all of the materials (like cement and plastic pipes) and paid for the labour. The project constructed a pipe network from a spring in Bukit Rigis to several concrete containers or tanks in the hamlets of Gunung Terang and Bedeng Sari. Connections from the tanks to the houses was not part of the government project, but was the villagers' responsibility. The plan did not materialise, the tanks were soon empty, and there was no supply of running water to houses in either hamlet. There were meetings to get the project done, but no concrete plan was decided on. One of the problems was the difficulty in getting agreement on a plan between the residents of the Bedeng Sari and Gunung Terang hamlets. One of the villagers noted that 'there were too many smart men in those two hamlets. Each insisted that his opinion was right and the others were wrong. They could come up with nothing!'

In contrast to the failure of the running water project in the hamlets of Gunung Terang and Bedeng Sari, the community in the hamlet of Rigis Bawah successfully carried out exactly the same project. Almost all of the residents have running water in their houses, and through weekly *gotong royong* they have built water tanks and installed a piping network throughout the hamlet. To ensure that every household participated in the weekly *gotong royong*, certain measures were agreed upon. Those who were absent from the *gotong royong* would either be prohibited from using the running water (channelling water from the tank to their houses) or obliged to pay a cash sum equal to a day's wage. The hamlet community found a clever way of obtaining cash to purchase the materials needed for the plumbing scheme. A few years previously, one of the hamlet residents had granted (*hibah, wakaf*) his coffee garden to the mosque in the hamlet. The accumulated profits from this garden were used to purchase the materials. The loan from the mosque was then paid back by each of those who enjoyed the running water.

Surprisingly, many residents of Rigis Bawah are also residents of both of the hamlets of Gunung Terang and Bedeng Sari where the water project failed. The problem of 'too many strong leaders, too much debate' in Bedeng Sari and Gunung Terang hamlets was often cited as the cause of the failure of the water project, but the fact that most of the houses in both hamlets had a well was also an important factor. The pressure of having the project done was high at the end of 2002, when wells were empty due to the long dry months in that year and their excessive use for watering the chilli gardens. Villagers with empty wells had started to go to the Way Besai River for bathing and washing clothes. There were meetings to discuss how to get the water project done, but as the discussions intensified, the rain came and the wells were filled. All talk on the water project subsided. In contrast, villagers in the higher altitude hamlets of Rigis Atas and Rigis Bawah still used springs and creeks as their primary water source. They had no wells, and failure to regulate water use created a serious crisis in these hamlets.

The people in Rigis Atas were now engaged in a community forestry contract. The head of the district granted them the right to farm in the state forest zone and they now had formal permission to use the land without worrying about being evicted or having their crops destroyed by the forest authority. It took two years for the community to arrive at a formal contract. The processes involved a detailed inventory, mapping, and formulation of rules and plans regarding the management of the cleared land and the remaining forest patch. The hamlet was involved in intense interactions with the field staff of WATALA and ICRAF who assisted them in the process, as well as local forestry officers. They were among the first of a small number of community groups in the province to engage in such a community forestry contract.

The cohesiveness of the village community was apparently contingent on need, urgency, resource availability or limitations, and finally leadership. The communal tasks performed by the villagers embraced development, religion and ritual, and the household economy. While extra-household relations played a role in villagers' livelihoods, most of the tasks in agricultural production were carried out by individual households. The next chapter explores this topic.

7. Social Stratification in Gunung Terang

Farming is the main occupation and source of income for most of the villagers of Gunung Terang. The few other occupations in the village include those of teacher, shopkeeper, reseller of farm produce, mechanic, builder, and car or motorbike taxi (*ojek*) driver. The proportion of villagers engaged in such off-farm work is relatively small — perhaps no more than 5 per cent. For most of those engaged in such activities, farming is still important, either as a primary or a secondary source of income.

When they were asked about the economic conditions of the families in the village, villagers often used the terms 'strong' (*kuat*), 'established' (*mapan*), and 'prosperous' (*makmur*) to refer to wealthy families; 'poor' (*miskin*), 'have not' (*tidak punya*), and 'needy' (*kurang*) for poor people; and 'enough' (*cukup*), 'not bad' (*lumayan*), 'ordinary' (*biasa*), 'common' (*kebanyakan*), 'average' (*rata-rata*), and 'on the edge' (*pas-pasan*) for those in between. The population of Gunung Terang village can thus be divided between a lower, medium, and upper stratum depending on their wealth. At the base are the poor, who comprise nearly half of the village population. The main characteristics of people in this group or class are their struggle to secure food to feed their families throughout the year, and their inability to afford their children's higher education. They usually live in huts (*gubuk*) or humble houses.

The middle stratum of the village population consists of those who worry less about feeding their family, but are more concerned with how to support their children's higher education, having a decent home, and possessing modern goods. The middle stratum can be further divided into what the villagers often refer to as *pas-pasan* (on the edge) or *cukup makan* (enough food), and *cukup* or *lumayan* (just enough, enough). While the former struggle to avoid becoming *kekurangan* (needy), the latter look for opportunities for further upward mobility. This stratum comprises about half of the village population. In the upper stratum, there are about two dozen families, roughly 3.3 per cent of a total of 708 village families or households, whom the villagers refer to as *mapan* (established) or *kuat* (strong). These families have successfully managed to accumulate wealth so that they have no problem feeding their families, building sturdy houses, or sending their children to university. They possess luxurious household goods and vehicles, and if they wish, they go on pilgrimages to Mecca.

The 'outer' hamlets of Rigis Atas, Rigis Bawah, Buluh Kapur, and Temiangan have a more or less equal number of low and medium stratum households. None of the wealthiest village families live there. In these hamlets, during 'good' years, the population tends to increase because of the arrival of new labourers, while it shrinks during poor years as the labourers move out. Medium stratum households dominate the 'inner' hamlets of Gunung Terang, Talang Jaya, Bedeng Sari, Petai Paya, and Simpang Tiga. All of the upper stratum households live there as well.

The following discussion illustrates the household circumstances of Gunung Terang villagers in the different economic strata. The stories are taken from fieldwork research in 2002, and some emphasis is placed on the processes of upward and downward mobility. Aliases are used for the names of individuals, but the places and times are real.

The Lower Stratum

Udin and his siblings were taken from Ciamis, in West Java, to Rigis Atas by their parents in 1973.[1] Udin was six years old then. His parents sold their small rice fields and upland field (*ladang*) in Ciamis. On their journey to Sumber Jaya, they were robbed in Kotabumi and lost all of their cash. Unable to buy land, the family cleared some state forest land near Rigis Atas and transformed it into 2.5 hectares of coffee gardens. Over the ensuing years, they bought a 1-hectare field within the state forest zone, which they then converted into a housing lot, 0.25 hectares of rice field, and a coffee garden.

In the 1980s, the forestry office commenced the reforestation projects and Udin's parents' gardens, along with hundreds of their neighbours' gardens, were planted with rosewood trees. They continued to look after and harvest the cherries from the coffee trees between the rosewood trees. In 1993–94, when the state forest zone's boundary was enforced, they were advised to dismantle their house and abandon their coffee gardens and rice field. After that, Udin, like his parents and siblings, started to work as a sharecropper in Gunung Terang and in the neighbouring village of Cipta Waras. In 2002, he was a sharecropper on 3 hectares of coffee gardens and was caring for a dozen goats. Udin, with the help of his brother, had also cleared a fallowed field belonging to the village schoolteacher and converted it into a rice field. He was granted the right to use the rice field (about 0.5 hectares) for two years. He, his wife and his six children lived in a hut in Bedeng Sari that belonged to the owner of the coffee gardens on which he was sharecropping. Although they did not go into debt to secure their food supply, the family had few possessions and could not afford to send their children for education beyond elementary school.

---

1   A few years earlier, Udin's father had visited Lampung to sell clothes and mats.

In working the gardens and rice fields, Udin and his wife regularly involved their younger siblings and, as as result, had to share any income with them. Udin and his wife regularly worked as wage labourers (*upahan*) and had a long-term plan to save their income to move back to West Java. A few years before, Udin had pawned a 0.25-hectare of rice field in his wife's village of origin in Bogor. The field was managed by his wife's parents. He wished to be able to save enough money to take over this rice field and move back.

Karya, in his mid-thirties, migrated to Rigis Atas with his parents and siblings in 1982. Following an eruption of Mount Galunggung that year, they vacated their house and left all of their possessions in their home village in Ciamis, West Java. Bringing nothing other than their clothes and kitchen utensils, wage labour was their primary source of income upon their arrival in Rigis Atas. There they cleared the forest, planted coffee, and built a decent house. The gardens and house were within the state forest zone, but in 1993–94, following the eviction of the 'forest encroachers' (*perambah hutan*), they abandoned their gardens and house. The family first moved to Banding, near Lake Ranau, for a couple of years. Later, they moved again to Simpang Luas, near Liwa. Their efforts to establish a new life in these new places were not as successful as they had been in Rigis Atas. Luckily, the family had managed to buy a housing lot near the hamlet of Gunung Terang. Karya, his wife, and two little daughters lived in a humble stilted house on this lot. His main sources of income were from a motorbike taxi (*ojek*) and from buying produce from Rigis Atas, such as bananas and chilli, sometimes jackfruit and avocado, which he took by motorbike to sell to middlemen in Fajar Bulan. He also cultivated a few hundred capsicum chilli in his small house garden.

Kamino, also in his mid-thirties, was considered by his neighbours to be one of the poorest people in the hamlet of Rigis Atas. His grandparents took care of him and his siblings in Ponorogo, East Java, when his parents joined a transmigration program to Rumbia, Central Lampung, in 1973. Kamino arrived in Lampung in 1989 when his parents had already moved to Mesuji, another transmigration site in the northern Lampung lowlands. In 1990 he came to Rigis Atas, cleared the bush, and planted coffee while working as a wage labourer. He sold his 3.5 hectares of coffee gardens, used the money to marry a woman from his parents' village in Mesuji, and bought a small plot of land. In 1993, after failing to make a decent life in Mesuji, he took his wife back to Rigis Atas and worked a hectare of coffee garden as a contract labourer (*bujang*), for which he was paid annually with a fixed amount of the harvest. The family had three small children and his eldest son was just starting to go to elementary school. Kamino and his family lived in a simple hut (*gubuk*) that belonged to the owner of the coffee garden that he was sharecropping. He also sharecropped another 1.5 hectares of young, non-bearing coffee garden belonging to another neighbour. Kamino had recently planted green beans on 1 *rante* (400 square metres) of unused land that he borrowed (*numpang*) from another neighbour. In 1999, he used all of his family's savings to make a down payment for a plot of coffee garden, but due to the drop in coffee prices, he was unable to complete the payments during the following years. As a result, the owner of the garden took the garden back without returning his down payment. Kamino and his wife obtained food for their family primarily from wage labour (*upahan*). Their income was so low that they could not even afford to buy government-subsidised 'poor rice' (*beras miskin*), the price of which is half of the market price but has to be paid for in cash. Kamino was well known in the neighbourhood as a strong and diligent man, but according to some of his neighbours, he did not manage his income well and he did not allow his wife to manage their finances, which was part of the reason for his failure to attain a better life.

Plate 7-1: A motorbike taxi (*ojek*).

Source: Courtesy of the author.

Hambali was the head of another poor family in Rigis Atas. In his fifties, with four teenage children, he and his wife struggled to pay the debts that they incurred to buy rice. He migrated to Lampung from Salatiga, Central Java, in 1979. He first lived in Simpang Sender, near Lake Ranau, working as a wage labourer in a coffee garden. In the mid-1980s, he and his family moved to Rigis Atas where they bought a fallow field and planted it with coffee while continuing to work as wage labourers. In 1993, they sold this plot and bought another 3 hectares of coffee garden, but soon sold the garden to pay their accumulated debts. Thereafter, they sharecropped (*maro*) a coffee garden and worked as wage labourers (*upahan*). Most often the wages that they received were much less than the debts they already owed. In 2001, they bought nearly a hectare of land covered with imperata grass, where they built a hut to live in and planted the rest with coffee. Hambali's son dropped out of junior high school, while none of his three daughters were educated beyond elementary school. His eldest daughter, who was 16 years old, had just started to work as a domestic helper in Jakarta. Hambali and his wife expected their other children to follow suit.

Like Hambali, Ahmadi — a Semendonese man in his mid-forties — was struggling to feed his family. He was no longer able to support his two sons studying at the junior high school, which forced them to drop out, and only his youngest daughter was still studying at elementary school. His wife was a *tunggu tubang*. She inherited her parents' house, a 0.6-hectare coffee garden, and a 0.5-hectare rice field, all of which were located in his wife's village of Srimenanti. His wife's parents were both sick and in constant need of his wife's care and cash for medication. The rice field was rented out to pay for his wife's parents' food and medical treatments. Ahmadi and his two sons lived in a hut in Rigis Atas and took care of a sharecropped young coffee garden. They regularly returned home to their house in Srimenanti. The land that they planted with coffee in Rigis Atas was his eldest sister's *tunggu tubang* property and was inherited from their parents. The land, about a hectare in size, was originally a productive terraced rice field which was abandoned in the 1980s when the reforestation project planted rosewood trees on it. The land was soon transformed into bush. In 2001, Ahmadi's family cleared the bush and planted it with coffee. According to Ahmadi, with any further drop in the coffee price, his family's investment in the coffee garden would have been a waste. Ahmadi was thus uncertain about the future of his family.

Plate 7-2: Harvesting coffee.

Source: Courtesy of the author.

Bi Ati, in her fifties, along with her husband and four children, had migrated to Lampung from Karawang, one of West Java's lowland rice bowls, in 1982. They first lived in Dwikora, a village on the eastern tip of Sumber Jaya, working as wage labourers, clearing the forest, planting coffee, and weeding gardens. In 1994–95, after the military operations to evict farmers from the state forest in Dwikora, the family moved around within Sumber Jaya before finally arriving in Rigis Atas. Here they sharecropped nearly 0.75 hectares of rice field and 2.5 hectares of an old, unproductive coffee garden belonging to a Semendonese villager who lived in Fajar Bulan. Bi Ati and her family lived in a house on stilts in the middle of the rice field. The income they received from the rice field was never enough to feed the whole family. According to Bi Ati, the low productivity of the rice field was largely due to a combination of low quality seeds, lack of chemical fertiliser, rat infestation, and poor upkeep. Rather than pouring all available labour into the rice field, the family frequently engaged in wage labour to pay the debts they incurred to buy rice. Warsi, Bi Ati's eldest daughter, had been working in Saudi Arabia since 2001, leaving her only daughter with Bi Ati, while her husband lived in Karawang. Asih, Bi Ati's second daughter, and Asih's small son were also living with Bi Ati. Asih's husband had just left her, and no one knew his whereabouts. Discussing her situation, Bi Ati once said:

I want my family to move back to Karawang. Being poor but close to relatives (*saudara*) would be better. Being poor without [having] anyone to turn to for help, like we are now here, is very difficult.

As soon as they had enough money to buy the bus tickets, Bi Ati insisted they would definitely return to Karawang and leave the region for good. But even saving money for bus tickets was difficult for the family.

Bi Ati's only son, Satria, in his thirties, was expecting his second child. He and his family had recently moved to a small hut belonging to the owner of 1.5 hectares of coffee garden that Satria was sharecropping. To buy rice for his family, he worked as a wage labourer and planted small capsicum chillies in the coffee garden. He endorsed his parents' decision to move back to Karawang as soon as possible, and said that if things got worse, he would follow in his parents' footsteps and return to Java for good with his family.

Ujang and his wife arrived in Gunung Terang in 1980. He was born in Gunung Terang hamlet, but since he was a boy had lived in Talang Padang, a Semendonese region in the neighbouring district of Tanggamus. He studied and married in Talang Padang. His wife was not a *tunggu tubang* so she did not inherit any of her parents' property. In Gunung Terang, Ujang taught young children to read the Qur'an. Initially, he received 15 kg of rice and 15 kg of dry coffee beans as an annual tuition fee from each of his pupils, but after 2000, none of his pupils' parents could afford to pay the fee. He lived in a simple stilted house belonging to his close kin. From 1995 to 1999, he was able to rent 1.5 hectares of coffee garden in Gunung Terang, and was sharecropping this garden in 2002. Ujang put a high priority on his children's education. His eldest daughter was a high school graduate and taught in an elementary school on a casual basis. His son and other daughter went to junior high school. With a very small income, his family could only support their children's education by maintaining a very simple life.

In addition to young families and households struggling for upward mobility, the lower stratum of the village was also occupied by older couples, widows, and widowers, many of whom were sick. Their children either lived elsewhere, or if they lived nearby they did not 'have enough'. While upward social mobility in the future was thought to be possible for the younger generation, it would be difficult if not impossible for the older generation.

The Middle Stratum

Triman, who was in his early fifties at the time of my fieldwork, departed from Salatiga in Central Java in 1978 and moved to Bukit Kemuning, where he first lived as a wage labourer. The following year, he used the savings he had accumulated to buy a 0.75-hectare coffee garden in Gunung Terang. He married a Javanese woman from Bedeng Sari, bought a housing lot, and built a simple wooden house in his wife's hamlet. They had eight children, none of whom had received an education beyond elementary school. Half of his children were teenagers who helped with his daily farming activities. Besides the coffee garden, he owned a 0.25-hectare rice field and half a dozen goats. The family cleared a fallow field belonging to a Semendonese villager, transformed it into a rice field, and were given the right to farm it for two years. As far as food security was concerned, Triman's family was in a better situation than those in the lower stratum, but the family still had to struggle to meet their other needs.

Plate 7-3. Hoeing a rice field.

Source: Courtesy of the author.

Ali was a Semendonese villager in his late thirties who lived in the hamlet of Gunung Terang. He was born in the village and married a Semendo neighbour. The family had two daughters, one in elementary school and the other in junior high school. As a *tunggu tubang*, his wife inherited all of her parents' property — a house, a 0.5-hectare rice field, and a 2.5-hectare coffee and pepper garden. Her parents were still able to feed themselves by farming a 0.8-hectare coffee garden, and they lived in a separate house (*turun*) located in that garden. The *tunggu tubang* rice field had been borrowed by one of Ali's wife's younger brothers, and its harvest was shared between Ali's wife, Ali's wife's brother who farmed the field, and Ali's wife's parents. Ali regularly hired labourers to weed and harvest the coffee and pepper garden. He and his wife did the other farming work themselves.

Syafri, in his early fifties, was born in Gunung Terang. In 1979, he married a widow with one daughter from her first marriage. The couple then had two more daughters and a son. His wife was a *tunggu tubang* in Muara Enim, in South Sumatra, and was entitled to the harvests of her parents' rice field and coffee garden. In 1980, Syafri bought 2 hectares of coffee and pepper garden, and in 1988 he bought a 1-hectare rice field and a 1.5-hectare coffee garden. He also bought a housing lot in Gunung Terang hamlet and built a sturdy wooden house. The house had luxurious possessions within, such as a big television, satellite dish, stereo set and nice furniture. The family managed their gardens and rice field on their own, while labourers were hired for weeding the gardens and hoeing and transplanting the rice field. The family would have two *tunggu tubang* daughters. Syafri's wife's daughter from her first marriage would be entitled to all of Syafri wife's parents' *tunggu tubang* properties in Muara Enim, while Syafri's own eldest daughter would inherit all of the family properties in Gunung Terang. The former was living with her husband and baby in a hut in the rice field. The latter had just graduated from high school and was preparing to study at a university in the capital of the province. Syafri's younger children were studying at junior high school.

Like Ali and Syafri, Effendi, a man in his mid thirties, also married a Semendo woman. The couple had two sons who were studying at elementary school. The family lived in a small but sturdy stilted wooden house in Rigis Atas, close to their garden. Unlike Ali and Syafri's wives, Effendi's wife was not a *tunggu tubang*. This couple acquired all of the properties they had by their own efforts. Both were born and raised in Fajar Bulan. In 1990, soon after their marriage, they cleared 3 hectares of state forest in the neighbouring region of Simpang Luas. A hectare of the cleared field was planted with coffee and the rest was transformed into an upland rice field (*ladang padi*). In 1993, while maintaining their coffee garden in Simpang Luas, the couple sharecropped 1 hectare of coffee garden in the neighbouring village of Srimenanti. They rented a house and lived in Srimenenati where Effendi's wife opened a small stall (*warung*). In 1996, using the money from selling their garden in Simpang Luas and the savings they had accumulated, they bought 2 hectares of coffee garden and 1 hectare of imperata (grass) field in Rigis Atas which was later planted with coffee. Effendi sharecropped half of his coffee garden and managed the other half with his wife. He hired labourers to weed and harvest the coffee garden and planted capsicum chilli in his coffee garden together with fruit and timber trees. Effendi owned and operated a movable motorised coffee mill, and was very busy milling his neighbours' coffee beans during the coffee harvest seasons.

Sutisna, a man in his early fifties, came to Rigis Atas from Ciamis, West Java, in 1979. With three other men, he worked as a contract labourer (*bujang*) and maintained Sucipta's coffee gardens in Rigis Atas. Sucipta was a trader who sold clothes from Tasikmalaya, in West Java, to various places in Lampung and South Sumatra. He bought 6 hectares of coffee gardens in Rigis Atas which were all managed by contract labourers he brought from Java. After his gardens delivered a peak harvest (*agung*), he sold the gardens and opened a clothing shop in Tasikmalaya. In the years that followed, Sutisna took over 2 hectares of Sucipta's coffee gardens and paid for them in three instalments — one each harvest season. In the early 1980s, following the drop in the coffee prices, Sutisna went to Palembang where he worked as a labourer in a chilli garden and a brick factory for a year. He returned to Rigis Atas and married a Javanese woman from the neighbouring village of Gedung Surian. He sold 1 hectare of his coffee garden and built a simple house. While maintaining the remaining hectare of coffee garden with his wife, he worked as a wage labourer and ran a motorised portable coffee mill. His wife opened a small stall (*warung*) selling items such as rice, cooking oil, sugar, salt, *micin* (monosodium glutamate), instant noodles, soap, cigarettes, snacks, and lollies.

In the mid-1990s, Sutisna sold his garden and house, and sent his wife and four children to live with his mother in Ciamis. He bought half a hectare of coffee garden nearby and built a hut to live in. He later bought two more coffee gardens, with a combined area of 1 hectare, in the nearby state forest. These had been abandoned by their owners after they were evicted by the military and forestry officers. In 2001, Sutisna planted capsicum chilli under the coffee and rosewood trees in his gardens. He was the first person in Rigis Atas to plant chilli for commercial purposes, and his neighbours frequently consulted him on how to plant chilli in their own coffee gardens. Sutisna regularly hired his fellow neighbours and his younger brother (who lived with him) as wage labourers to manage his coffee and chilli gardens. With the income from his chilli plants he was able send cash to his family in Java on a regular basis.

Plate 7-4: Harvesting chilli.

Source: Courtesy of the author.

The Upper Stratum

Fahrozi, a Semendonese villager in his mid-forties, was born in Gunung Terang. He married a *tunggu tubang* woman who inherited a big wooden stilted house, 4 hectares of coffee and pepper gardens, and 2.5 hectares of rice fields. The family managed 1 hectare of the garden, while sharecropping the rest of the garden and the rice field. His wife's parents moved out (*turun*) of the house to occupy and manage coffee and pepper gardens elsewhere. Until 1999, Fahrozi was active in the coffee and pepper reselling business, and was one of half a dozen coffee middlemen in the village. In 1999, he bought 10 hectares of bush land in the neighbouring region of Sukau. When the coffee price fell, he did not have enough capital to carry out his plan to plant his fields with coffee and pepper, but he was still able to afford to build a big sturdy wooden stilted house as a family home. Fahrozi had three children. His eldest daughter had just graduated from high school and was preparing to study further in the provincial capital, Bandar Lampung. She would receive the house and other properties in her *tunggu tubang* capacity. Fahrozi believed that his sons — the first then studying at elementary school and the second still under school age — would also go to university. Otherwise, they would inherit the bushland that he had just bought and become farmers.

Sunaryo was a Javanese villager in his sixties who came to Gunung Terang in 1983. Prior to that, he and his family had come from Purwodadi in Central Java to join a transmigration program in Rumbia, in lowland central Lampung, in 1974. The lack of irrigation in Rumbia forced the family to leave the transmigration site. In Gunung Terang, Sunaryo began his business cutting and selling timber from the state forest in Rigis Atas where the family first lived. He ran the business for over a decade with the backing of the local police and military officers, without whose support he would have been jailed. Sunaryo used the proceeds from the timber business to buy 4 hectares of old coffee garden in 1983, 2 hectares in 1987, 1 hectare in 1992, and 2.5 hectares in 1997. In 1997, he also bought 1.5 hectares of rice fields which he soon converted to a coffee garden, which meant that he had a total of 11 hectares of coffee gardens. In the same year, he bought a 0.25-hectare housing lot (*kapling*) in Bedeng Sari, built a large brick house and moved into it. Sunaryo had seven children, of whom three — a daughter and two sons — were already married. Sunaryo gave each of these three children 1.5 hectares of coffee garden and a house. His other four sons, all in their twenties and either junior or senior high school graduates, collectively managed the remaining 6.5 hectares. Each of them would inherit the same area of land when they got married. Sunaryo and his wife said that they would bequeath the house and housing lot to their youngest son or the last one to marry, who would in return take care of them in their old age. The family had been cultivating red chilli in their housing lot and small chilli in the coffee gardens.

Unlike Fahrozi and Sunaryo, Sabar and Rahman had much less land, but much more wealth. Sabar had only 3 hectares of coffee garden but was an active coffee middleman and, more importantly, was a moneylender. His family lived in the capital of the province, Bandar Lampung. His house in Gunung Terang functioned more as a store for sacks of dried coffee beans which he bought and resold, and an office for his moneylending business. Rahman had no coffee gardens, but he did own the largest shop (*warung*) in the village that sold household items. Following in Sabar's footsteps, Rahman also engaged in moneylending in the village. Over a decade before, after some years of work as a *kenek* (bus driver's assistant), he and his wife rented a small house and opened a small shop. He started his moneylending business as a broker, and later set up his own service. He was the most active moneylender in the village in 2001–02.

Two village officials — Bu Mas Muda, the village head, and Udin, the chair of Lembaga Pemberdayaan Masyarakat Pekon (LPMP), the village council for community empowerment — were among the established (*mapan*) families in the village. According to villagers, it was not because of their official position that they became wealthy. On the contrary, the fact that both were in established families was the main reason that the villagers chose them as village officials. Both families owned more than 3 hectares of highly productive coffee gardens and had also started commercial vegetable farming.

As far as wealth is concerned, the wealthy families in Gunung Terang were much less wealthy than the rich merchants (big coffee resellers and owners of big retail shops) in Fajar Bulan, Sumber Jaya and in other villages. For these rich merchants, the amount of land owned is not the determining factor for wealth accumulation. Access to capital and trade networks matter more.

8. The Farming Economy in Gunung Terang

One of the key factors observed in relation to socio-economic differentiation in the village was ownership or control of the land. Poor villagers were landless or nearly landless, and included recent migrants and younger generation villagers who had not yet inherited their parents' land. They cultivated other people's fields or possessed only a small plot of land. A few of them owned or controlled a relatively large area of land but did not have the capital to develop productive cultivation. While some wealthy households controlled extensive areas of cultivated land, land ownership in itself was not a factor in household economic status. Two households owning or controlling the same area of land could belong to different economic strata.

The smallholding farmers in Gunung Terang were rational, flexible, and responsive to constraints and opportunities. In the years of attractive coffee prices, they poured labour and capital into coffee farming and adopted modern techniques in the process. Following the recent drop in coffee prices, they responded by reducing labour and non-labour inputs in coffee farming and investing elsewhere. The result was a sharp decrease in the return to land for coffee production, but returns on labour remained almost as high as the return on labour from small chilli cultivation. In contrast, more labour and non-labour inputs were directed to vegetable farming, where productivity — in terms of the return both to land and to labour — was higher than it was from the cultivation of both coffee and rice. This reluctance to wholly abandon coffee production in favour of the higher return from vegetables was based on the advantages of strategic diversification and the opportunity to intensify coffee production in the future when prices might increase again.

Smallholder production of agricultural commodities for domestic and global markets was also made possible because of significant production inputs obtained through non-market relations. This was reflected in patterns of mutual assistance and reciprocal labour exchange among kin and neighbours for both food and commodity production. Access to productive land could also be acquired through inheritance, sharecropping and borrowing (*numpang*) arrangements, or simply by squatting on forestry land. Forms of usury (*musiman*), revolving credit among neighbours (*arisan*), and interest-free loans from friends and relatives provided alternative sources of capital. The ability to discount a range of production costs by engaging social capital enabled smallholders to profit on the margins of commodity production.

Household Farming Practices

Coffee gardens dominated land usage in Gunung Terang village, just as it did in other villages in the region. According to elders in the village, leaving old coffee gardens fallow, and clearing the forest or old fallow for new gardens, was a common practice until the early 1980s. Thanks to the introduction of subsidised chemical fertilisers and better farming techniques (pruning and grafting, weeding and soil conservation) adopted as a result of various agricultural extension programs, the system of rotational coffee cultivation has not been practised since the 1980s. With the new techniques, the productivity of coffee gardens had risen dramatically. Under the old rotational system, a hectare of coffee garden would produce 0.7 tonnes per annum during the peak (*agung*) stage, and an average of 0.3 tonnes for the remaining years until the plants ceased to bear fruit at the age of 10 years or more, at which time the field would be left fallow. Under the newer system, a hectare of coffee garden would normally produce 1.5 to 2 tonnes of dry coffee beans each year over the same length of time. An attractive coffee price and the arrival of new migrants enabled the success of the newer coffee farming system.

Plate 8-1: Weeding a coffee garden.

Source: Courtesy of the author.

In the early 2000s, there was a decline in coffee garden production. As Table 8-1 shows, the average production in 2002 was three times lower than in 1998. The decline in production from 1998 to 2002 coincided with a drop in the coffee price from between Rp 12,000 and Rp 15,000 per kilogram in 1998, to between Rp 6,000 and Rp 7,000 in 1999, between Rp 3,000 and Rp, 4,000 in 2000, and only Rp 3,000 in 2001 and 2002. In response to the fall in the coffee price, non-labour inputs (fertiliser) and/or labour inputs (weeding and pruning) were gradually reduced, which led to a further drop in coffee garden production. Production tends to fall as the price drops, and local smallholders argued that a higher price in the future would revive production.

Table 8-1: Coffee garden production in Gunang Terang village, 1998–2002.

Farmer	Garden size (ha)	No. of trees	Description	Annual production (tonnes per hectare)				
				1998	1999	2000	2001	2002
A	2.0	5,000	Grafted, 0.25 tonne fertiliser applied in all years	1.65	0.75	0.65	0.75	0.62
B	1.6	4,000	Grafted, 0.6 tonne fertiliser applied in 1998	1.85	1.85	0.45	0.45	0.45
C[a]	1.0	2,500	Grafted, 1.0 tonne fertiliser applied in 1998	2.3	1.7	0.8	0.7	0.7
D[a]	0.75	1,800	Non-grafted, 0.6 tonne fertiliser applied in 1998–2000	3.3	1.86	1.7	1.7	1.6
E	2.6	6,500	Grafted, inter-planted with pepper,[b] no fertiliser applied all years	0.77	0.65	0.34	0.3	0.2
F	2.0	5,000	Grafted, inter-planted with pepper,[c] no fertiliser applied all years	3.5	2.5	2.0	1.75	1.75
G	1.4	3,500	Grafted, inter-planted with pepper,[c] no fertiliser applied all years	2.5	1.25	0.5	0.5	0.5
H	1.45	3,600	Grafted, inter-planted with pepper,[d] 0.5 tonne fertiliser applied all years	2.5	0.67	0.67	0.67	0.2
Average	1.62	3,988		2.30	1.40	0.89	0.85	0.75

Notes: (a) farmers C and D were sharecroppers; (b) farmer E had 1,000 vines, producing 2,700 kg in 1998, none from 1999 to 2001, and 6 kg in 2002; (c) farmers F and G had 100 to 200 young vines; (d) farmer H had 1,500 vines producing 600 kg in 1998 and 1999, and none from 2000 to 2002.

Source: Interviews with villagers in Gunung Terang, 2002.

Table 8-2: Inputs and income from a 1-hectare coffee garden, 2002.

Inputs		Annual average
Labour (weeding)	32–150 person days	54
Labour (pruning)	7.5–60 person days	25
Labour (harvesting)	12.5–58 person days	21
Labour (drying)	7–14 person days	12.5
Total labour inputs		112.5 = A
Non-labour inputs (milling)	Rp 150 x 700 kg	Rp 105, 000 = B
Income		**Annual average**
Gross income	Rp 3,000 x 700 kg	Rp 2,100,000 = C
Net to landowner hiring labourers	C − (A x Rp 10,000) − B	= Rp 870,000
Net to sharecropper/landowner	0.5 x (C − B)	Rp 997,500 = D
• As return to labour	D ÷ A	= Rp 8,867
• As income in milled rice	D ÷ Rp 2,000 per kg	= 499 kg

Source: Interviews with half a dozen villagers in Gunung Terang, 2002.

Despite the drop in the price and production of coffee, coffee gardens were still seen as an important household income source. Table 8-2 indicates that a hectare of coffee garden producing 0.7 tonnes of coffee provided modest but significant income to the household economy. Regarding the return to labour, smallholders received a daily income approximately equal to the daily wage rate in the region — Rp 10,000 per day. The equivalent rice yield would be 0.5 tonnes of milled rice, and that, according to some villagers, would be just enough to feed a small family for a year. Milled rice, however, is only one item among many other basic household needs. Even to have an income worth double the minimum stock of milled rice is considered to be marginal or 'on the edge' (*pas-pasan*). With the low coffee prices and production, low incomes became a problem with which farming households struggled to cope.

Although dominant, coffee was rarely the only crop planted in the gardens. Many coffee gardens had shade trees (*dadap* and *ki hujan*) that also supported pepper vines. However, pepper had recently become a less important cash crop as it was prone to various diseases. The gardens closer to housing lots had a greater variety of crops. Fruit trees such as coconut, jackfruit, avocado, rambutan, and *jambu* (guava) were planted, but only for domestic consumption. Bananas also become a commercial commodity, and various other fruit trees and timber trees were planted, though their economic importance was unclear. Vegetables, spices, and tubers (cassava and sweet potato) — all for household consumption — were annual crops easily grown in the kitchen gardens of villagers' houses. Stall-fed sheep and goats also emerged as part of the village economy. The demand for compost for commercial vegetable farming drove the emergence of livestock husbandry in the village.

Unlike commercial vegetable farming (*kebun sayuran*), rice fields were always important. From a total of more than 1,500 hectares of village land, about 90 hectares were rice fields. Some rice fields had been abandoned or converted into coffee gardens during the last peak in coffee prices in the mid-1990s, but many of the abandoned fields were later restored to rice production. Interviews with a dozen of the farmers who were farming the rice fields revealed that half of them were landowners while the other half were sharecroppers. Field size varied from 0.25 to 1.5 hectares (with an average of 0.725 hectares), and output varied from 0.485 to 2.1 tonnes per hectare (with an average of 1.3 tonnes). With normally two crops per year, households who farmed a rice field had relatively greater food security.

Table 8-3: Inputs and production from a 1-hectare rice field, 2002.

Inputs	Annual average	Annual cost
Labour (hoeing and ploughing)	28 person days	Rp 400,000
Labour (seed bed)	0.5 person days	
Labour (transplanting)	27 person days	Rp 300,000
Labour (fertilising)	4 person days	
Labour (weeding)	68 person days	Rp 275,000
Labour (spraying)	2 person days	
Labour (harvesting)	24 person days	
Total labour inputs	153.5 person days	
Non-labour inputs (fertiliser)	300 kg x Rp 150	Rp 450,000
Non-labour inputs (pesticide)	4.5 lt x Rp 40,000	Rp 180,000
Total cost of inputs		Rp 1,605,000
Production		
Total harvest	1,750 kg	
• Harvesters' share	250 kg	
• Net return	1,500 kg	

Note: labour for winnowing and drying, and costs for seeds, transportation, and milling are excluded.

Source: Interview with a villager in Gunung Terang, 2002.

Table 8-3 illustrates the input and production of a hectare of rice field. According to those who farmed rice fields, production could be increased further to at least double the level shown in this table. However, there were constraints to such improvements in production. Lower or late rainfall would reduce rice field production, and too much rain would flood rice fields on riverbanks and destroy the harvest. Fungus, insects, and pests (mainly rats but also pigs in fields close to forests) were constant threats, and seed was always a problem. Local rice varieties (*belebur rimba*) produced well but needed five to seven months to ripen. Seeds of high-yielding varieties (like Cisadane or IR 36) were not readily available in local shops, so seeds from the harvest were used instead. Credit for fertiliser was unavailable, and the availability of extra labour for proper weeding and repair of bunds, terraces, and ditches was another constraint, particularly for households who also farmed coffee and/or regularly engaged in wage labour. The labour shortage in the rice fields always limited the achievement of higher outputs.

All villagers who farmed rice fields in the village also farmed coffee, ensuring their household income and food security. Since 2001, chilli had become another important commodity. There were two types of chilli farmed in the village — hot small capsicum (*cabe kecil, lombok*) and big red capsicum (*cabe merah, cabe besar*). Several varieties of each type of chilli were planted in the village. The main difference between small and big capsicum is that the former was planted under the coffee trees in the coffee garden, while the latter was planted in open fields. Small capsicum farming was the favourite choice of households in the lower strata because it promised modest profits but needed little in the way of non-labour inputs, while red capsicum farming, which required considerable cash investment but promised a more lucrative return, was only practised by those in the medium and upper strata. Tables 8-4 and 8-5 illustrate the inputs, production and income for both types of farming. At the end of 2002, less than 50 villagers could afford to farm red capsicum, while more than 100 households cultivated small capsicum in their coffee gardens. A few of them had also experimented with other commercial vegetables such as beans, tomato, shallots, and eggplant in their house gardens, but chilli was still the favourite choice at the time.

It is possible to harvest small capsicum twice a month for up to 18 months. Maintaining small capsicum crops in coffee gardens saves labour, particularly on weeding, while the use of compost on the capsicum was also good for the coffee crop. Iman, a sharecropper on a coffee garden in Rigis Atas, claimed that his 1,300 small capsicum plants 'fed' his small family in 2001–02. The sacks of dry coffee beans he earned as his share were treated as the family's savings. The number of plants was a critical factor in determining the amount of family labour needed to maintain the crops. Up to 1,500 small capsicum plants could be easily managed by a couple, but beyond that number hired labour would be required.

Table 8-4: Inputs and income from 1500 small chilli plants, 2002.[a]

Inputs	Annual average	
Labour (land preparation)	10 person days	
Labour (seedlings)	3 person days	
Labour (transplanting)	2 person days	
Labour (weeding)	25 person days	
Labour (compost application)	5 person days	
Labour (spraying)	2.5 person days	
Labour (harvesting)	80 person days	Rp 400,000 labour hire = A[b]
Total labour inputs	127.50 person days = B	
Non-labour inputs (compost)	10 x 50kg sacks x Rp 10,000	Rp 100,000
Non-labour inputs (pesticide)	0.5 litre @ Rp 20.000 per litre	Rp 10,000
Total cost of non-labour inputs		Rp 110,000 = C

Income		
Gross income	750 kg x Rp 2,000	Rp 1,500,000 = D
Net income in cash	D – (A + C)	Rp 990,000 = E
Net income in rice	E ÷ Rp 2,000 per kg	495 kg
Return to labour	(D – C) ÷ B	Rp 10,902

Notes: (a) in an open field 1500 chilli plants required a *rante* (400 m2) of land, but when inter-planted in a coffee garden, chilli can be planted in equal numbers with the coffee trees; (b) half of harvest labourers were hired (hence 40 person x Rp 10,000).

Source: Interview with 2 villagers in Rigis Atas, 2002.

Unlike small chilli that can be planted under the coffee trees, red chilli could not tolerate shade, which meant that coffee gardens had to be converted in order for the chilli to grow. Most households that planted red chilli converted 2 *rante* (800 m2) of their coffee gardens. Less than a dozen households could convert a larger area. As in the case of small chilli, an area bigger than this would require hired labour. Unlike small chilli, which can last up to 18 months, a cycle of red chilli lasted only five months. According to most villagers, a lack of adequate capital, declining soil fertility, and the risk of disease were critical factors that hindered the sustainability of red chilli farming. The lucrative return promised from red chilli farming was believed to be short-lived. Many villagers predicted that planting red chilli three years in a row would result in *tanah mati* ('dead soil'), which meant that no valuable crops would then grow in it. To make matters worse, diseases caused by resistant viruses and/or fungus could no longer be cured, leaving villages with little option other than to leave the field fallow or to plant trees on it. In order to prevent this, some villagers felt it was urgent to experiment with other commercial vegetables that were planted rotationally. Some villagers speculated that if coffee prices remained low, smallholders in the village and the region would turn to commercial vegetable farming. It was not clear how smallholders in the village would succeed in coping with such market dynamics and ecological limitations.

Table 8-5: Inputs and income from 2 *rante* of 'big' red chilli, 2001.[a]

Inputs	Annual average	
Labour (land preparation)b	90 person days	Rp 900,000
Labour (seedlings)	7 person days	
Labour (transplanting)	2 person days	
Labour (making and putting sticks)	6 person days	
Labour (compost application)	19 person days	
Labour (spraying)	0.5 person days	
Labour (harvesting)	60 person days	Rp 600,000
Total labour inputs	184.5person days	
Non-labour inputs (fertiliser)	300 kg	Rp 375,000
Non-labour inputs (compost)	60 sacks x Rp 5,000	Rp 300,000
Non-labour inputs (fungicide)	1 kg	Rp 46,000
Non-labour inputs (pesticide)	0.5 litre @ Rp 70.000 per litre	Rp 35,000
Non-labour inputs (plastic mulch)	20 kg x Rp 13,000	Rp 260,000
Non-labour inputs bamboo	20 x Rp 3,000	Rp 60,000
Total cost of inputs		**Rp 2,576,000 = A**
Gross income	**3,000 kg x Rp 5,000**	**Rp 15,000,000 = B**
Net income	**B – A**	**Rp 12,424,000**

Notes: (a) a *rante* (400 m2) of open field can be planted with 800 chilli plants; (b) this included uprooting the coffee trees.

Source: Interview with 3 villagers in Gunung Terang, 2002.

Land, Labour and Capital

Like smallholders in most hilly areas in Lampung, villagers in this area recognised two types of land, *tanah kawasan* and *tanah marga*. The former term refers to officially designated state forest zones (*kawasan hutan negara*) which are also called BW (from the Dutch *boschwezen*). The second term refers to non-state land that can be individually owned and is eligible for land title (*sertifikat tanah*). There are different terms used for buying and selling these two types of land. *Jual* (selling) or *beli* (buying) are terms used for transactions in *tanah marga*, while *ganti rugi* (compensation) is used for transactions in *tanah kawasan*. Unlike *tanah marga*, where individual ownership is secure, *tanah kawasan* always carries a risk of crop destruction, confiscation, and eviction.[1]

---

1   This is why smallholders in Rigis Atas were so keen to engage in the HKm community forestry contract.

The price of a coffee garden in *tanah marga* was more than twice the price of one in *tanah kawasan*. For example, in 1999–2000, one hectare of productive coffee garden in *tanah marga* in Bedeng Sari was priced between Rp 20 million to Rp 40 million, while at the same time a *kawasan* in Rigis Atas would cost less than Rp 10 million per hectare. In 2002–03, with the drop in the coffee price, the price of land in both *marga* and *kawasan* declined to roughly half of these prices. In practice, the number of coffee trees matters more than the area of land. Normally, 2,500 coffee trees grow on 1 hectare of land, so the size of the trees determines the price.

While ownership of a plot of *belukar* (fallow or bushland) in *tanah marga* is secure, in *kawasan* it is otherwise. There are stories of smallholders whose abandoned gardens in *kawasan*, which had been simply left during the eviction operations and turned into *belukar* in the following years, were taken or sold by someone else. There are also cases of smallholders who were given a *belukar* plot for free by fellow villagers who were giving up farming *kawasan* land. Transactions in coffee gardens could generally be achieved with a down payment followed by up to three annual payments (*cicilan*) after the harvest season. There are examples of villagers who have lost their gardens because they failed to complete the payments due to poor coffee harvests and/or low prices. In these cases, the family would lose all of the money that had already been paid.

Due to the recent drop in coffee prices, the practice of renting a coffee garden had understandably become more infrequent. For example, Ujang rented a 2.5-hectare coffee and pepper garden for Rp 3 million in Gunung Terang hamlet for seven years until 1999, when he changed the tenancy to sharecropping in anticipation of a further drop in the coffee price. Renting rice fields was more common. For example, Ahmadi's wife had for years rented out her one hectare *tunggu tubang* rice field in Rigis Atas for 100 *kaleng* of *gabah* (unhusked rice) from the first crop and 60 *kaleng* from the second.[2] A few villagers who farmed red chilli gardens planned to rent land in the future if the price of the red chilli remains stable.

Sharecropping (known as *maro* or *garap* in the case of coffee gardens and rice fields, or *njawat* in the case of rice fields) is a common way of combining land and labour in the village and the region as a whole. The 2001 village income statistics (*mata pencaharian*) recorded 808 adults working in agriculture. They included most of the male heads of the 706 households in the village plus other adult males residing in the same houses. Women were excluded from these statistics. About 60 per cent of the villagers who specified agriculture as their main source of income were *petani* (owner cultivators), while 40 per cent were *buruh tani*

---

2 Depending of the quality of the rice, one tin container (*kaleng*) of 13–16 kg of *gabah* yields 5–8 kg of milled rice (*beras*).

(non landowners or agricultural labourers). Most *buruh tani* sharecropped land belonging to the *petani* while regularly engaging in wage labour. About half of these *buruh tani* farmed only coffee gardens, while the other half worked on both rice fields and coffee gardens.

There were variations in the sharecropping arrangements for coffee gardens and rice fields. For a coffee garden, the sharecropper was usually responsible for all of the labour inputs, namely the weeding and pruning. The harvest was shared equally with the landowner after other costs, like the purchase of fertiliser and the cost of harvest and milling, were deducted. There were also variations in the arrangements for these additional costs to be covered. The cost of fertiliser could be deducted from the total harvest or it could be the responsibility of either the landowner or the sharecropper. There were cases where labour for harvesting was the sharecropper's responsibility and the costs were deducted from the total harvest. There were also exceptional cases where the sharecropper received only one third of the harvest.

In the case of rice, the harvest was equally divided (*bagi dua, paro*) for the first crop (*tanam pertama, rendengan, musim tahun*), which is planted in the rainy season (the first months of the year) and harvested in July and August. But with production declining by up to one half in the second cropping period (*tanam kedua, gaduh, parekat, musim selang*), which starts in September, the harvest was often divided in three, with the landowner receiving one third and the sharecropper two thirds. The arrangement for labour and other costs was fixed. All of the labour — except for harvesting — was the sharecropper's responsibility. The additional costs of fertiliser and labour for harvesting were deducted from the total harvest.

A sharecropper could terminate a tenancy anytime that he or she wished, but a landowner could not. *Minggat* (leaving without saying) is a negative term that was applied to a sharecropper who terminated an arrangement without notice or only short notice. When the landowner wanted to farm his or her own garden, or choose someone else to sharecrop the garden, the sharecropper could ask to remain as a sharecropper of the garden for a couple of years even after the arrangement had ended. Although lending money or rice to the sharecropper was common, it was not an obligation on the part of the landowner, but was a way to prevent the sharecropper from *minggat*. *Numpang*, which meant free access to land in both rice fields and coffee gardens for housing, was common in sharecropping arrangements. In cases of small chilli inter-planted in a coffee garden, the sharecropper was entitled to all of the chilli harvest if the landowner provided no funds for the purchase of fertiliser and other necessities. Another version of *numpang* involved the right to use rice fields. By converting land (abandoned rice fields, coffee gardens, or bushland) into a rice field by levelling the field or building terraces and channelling the water, a family could 'own' the field for a year or two.

Wage labour *(upahan)* was an important source of income for households in the lower stratum in the village. Jobs that were done by wage labourers included: ploughing, hoeing *(macul)* and harvesting rice fields; weeding *(ngoret)*, pruning *(buang ranting, buang tunas/wiwil)* and handpicking *(mutil)* coffee cherries in coffee gardens; and more recently, land preparation and harvesting in chilli gardens. *Upahan* could be done on the basis of a daily wage *(harian)* or contract *(borongan)* for all jobs except the harvesting of rice. The daily wage rate in 2002 was Rp 10,000 without meals, or Rp 5,000–6,000 with three meals (food and coffee) plus cigarettes. *Borongan*, work would normally be completed faster and cost less than *harian* work. For example, weeding one hectare of coffee garden took 15 person days and cost Rp 150,000 in the *harian* system, but was done in no more than 10 days and so cost only Rp 100,000 in the *borongan* system. A common way for a man to get the job done in the *borongan* system was to enlist the aid of his wife and children to complete the job. Another way was by working harder and/or longer hours. A strong labourer could do the work quicker and earn more than Rp 10,000 per day. *Borongan* for the hoeing of rice fields, which required skill as well as strength and longer working hours, was the kind of job in which a daily wage higher than Rp 10,000 could be earned.

Giving milled rice in addition to cash payments used to be quite common in both *harian* and *borongan*. For example, in 2000, Kamino, a labourer in Rigis Atas, might earn either Rp 100,000 plus 25 kg of milled rice, or Rp 130,000 minus the rice, by working for 10 to 15 days weeding and pruning coffee gardens. After 2001, he found no one willing to include rice as partial payment of his wage.

There were also two different arrangements for harvesting rice fields. In the *derap* system, workers received one sixth or one seventh of the total harvest as their share *(bawon)*. In the *ngepak* or *ceblok* system, the workers were generally not paid at all, but were sometimes provided with meals and might still receive a share of the harvest. The second system frequently involved close kin, friends, or neighbours, and was becoming more common in the neighbouring villages with larger rice fields. Elders in Gunung Terang hamlet insisted that payment of wages for harvesting rice fields was a recent trend, and that in the old days reciprocal help *(bantu)* was the rule.

Wages for coffee harvesting could be paid either in cash on a daily basis or in kind, and harvesters received a share of one fourth to one fifth of the crop. In 'good' years, a worker could handpick *(mutil)* 10–15 *kaleng* of coffee cherries in a day, but in 'poor' years, only 2–5 *kaleng*.[3] A daily *harian* wage was always paid to workers handpicking both types of chilli.

---

3   One *kaleng* of coffee cherries weighing 15 kg yields 3 kg of dried milled coffee beans.

From the 1970s to the mid-1990s, annual contract labour (*bujang*) was common in the village and the region. The term *bujang* is used to refer to young unmarried male contract labourers.[4] In this arrangement, the *bujang* labourer was paid annually, either in cash or in kind (with sacks of coffee beans). For example, in the 1980s, a *bujang* working 2 hectares of coffee garden in Rigis Atas received a sum of cash equal to the price of 2 tonnes of milled rice. Moreover, the landowner was responsible for providing shelter and meals for the *bujang*. In the 1990s, some villagers had three or more *bujang* in their household. By 2002, none of the villagers had *bujang*.

Reciprocal labour exchange (*gantian, gentenan, liuran, royongan*) in coffee gardens was said to be common before the recent 'poor' years. Half a dozen or more men formed a temporary group and worked in each other's fields, weeding and pruning their coffee gardens. In 2001–02, the system was used by villagers to cultivate chilli in most of the hamlets, but there were only two or three people in each group. Regardless of the change in group size, reciprocal labour exchange was a practice confined to people in the lower social stratum of the village.

Failing to pay their debts under *musiman* (the local moneylending system) was a common way for villagers to lose their land, their house, or their crops. The land was simply taken over (*ditarik, dicabut*) by the moneylender if the debt was not paid. The interest rate on this kind of debt was between 70 and 100 per cent per year. After a few years, the accumulated debt would be close to the market price of the land, thus allowing the moneylender to sell it. This could be done with a land title (*sertifikat*) or a blank duty stamp paper (*kertas segel*) signed by the debtor.

Unlike *musiman*, a *gadai* (pawning) arrangement carried no risk of losing the land. With this type of arrangement, villagers received a certain amount of cash, no interest was charged, but the lender was entitled to all of the harvest until the debt was paid off in full. If debtors needed more money, they would seek a line of credit from the local branch of Bank Rakyat Indonesia (BRI). Land title as collateral was a prerequisite. According to some villagers, in the hamlet of Bedeng Sari alone, at least 50 households had received credit from BRI and all of them were from the medium and upper strata. For those in the lower stratum, the absence of land title and the high cost of transportation to get to and from the bank to secure the credit prevented them from obtaining it. During the 'bad' years, neither BRI credit nor *gadai* arrangements with fellow villagers were available, which meant that *musiman* became the only alternative. The high profit from red chilli cultivation (see Table 8-5) was seen as a way to offset this burden.

---

4   *Bujang* is a Malay vernacular term for single, unmarried man.

The returns from red chilli cultivation had changed the arrangements for payment of outstanding *musiman* debts. Rather than taking over the coffee garden, the moneylenders would turn the debtor into a red chilli garden sharecropper. The moneylender financed all of the start-up costs, while the debtor household was responsible for the day-to-day upkeep of the garden. The net income — total revenue minus start-up costs — was divided equally. However, the outstanding debt plus the accumulated interest remained intact. Apart from the profit promised by red chilli, the change in *musiman* arrangements was also due to pressure from villagers on village officials to prevent their land from being taken over by the moneylenders. With the drop in production and sales of coffee, more and more villagers would have had their land taken. One thing the village officials could do to prevent this was to refuse to put the village's official sign and stamp on the land transfer papers. To back the village officials, a few educated villagers also threatened to report the moneylenders to the police for practising usury, which by law was a crime.

According to villagers, the *musiman* debt system emerged at the same time as the expansion of coffee production in the region in the late 1970s and the early 1980s. It began with the infamous *satu dua* ('one two') debt. For example, one sack of coffee beans borrowed four to six months prior to the coffee harvest season would be repaid with two sacks, or a sack of milled rice borrowed during the planting season would be repaid with twice the amount four months later. Newly arrived migrants were the ones who usually engaged in *satu dua*, and coffee resellers, shopkeepers, and *haji kopi* were the likely sources of such loans. Later, to avoid losing their profit, the moneylenders demanded cash payments amounting to double the cash value of the coffee or rice when it was borrowed. Soon, the system of borrowing cash with 100 per cent interest per year was introduced, along with accumulation of debts from year to year. The increasing price of coffee from the late 1980s to late 1990s encouraged the *satu dua*, and later the *musiman*, systems in the region.

There were opposing views among villagers on the practice of *musiman*. Some perceived it as a sin (*dosa*) to accumulate wealth in this way, and referred to Islam's prohibition on usury (*riba*). Others saw it as a normal cash transaction, no different from obtaining a loan from a bank. The latter opinion was based on the fact that the usurers never openly offered their services, and were often the ones who were approached and needed to be convinced to participate. However, all agreed that it was immoral to derive large profits from *musiman*.

The hamlet of Bedeng Sari had two men running the *musiman* moneylending business — Rahman, of Ogan origin, and Samsi, who was Javanese. Both also ran retail businesses in their large shops (*warung*). There was a third moneylender in Simpang Tiga — Haji Sabar, a Semendonese who ran a coffee reselling business. Villagers in Gunung Terang were also the clients of two big moneylenders in the neighbouring village of Sumber Alam — Indra, an Ogan, and Barno, who was Javanese. Indra also had a retail shop, while Barno ran a coffee reselling store. Barno was the wealthiest moneylender, owning a luxurious two-storey house, with a large store attached to it, and some trucks and minibuses. An elder in Gunung Terang once made the exaggerated claim that the number of villagers' land certificates that Barno kept as collateral for *musiman* debts was equal to the number held by the BRI branch in Fajar Bulan as security for villagers' debts. When Barno died in the early 2000s, Indra replaced him as the biggest moneylender in Mutar Alam.

In general, villagers looked down on those who ran these businesses, and tried as hard as possible not to be trapped in *musiman* debt. Close kin, friends, and neighbours were the ones they turned to for help when there was no rice to cook or when a small amount of cash was needed to buy medicine or visit the health clinic. It was considered immoral not to lend cash or rice to a close friend, neighbour or relative when they were in dire need. It was reciprocal help and assistance that strengthened the bonds and cohesion among villagers. For those in the medium and upper strata, such mutual help among close kin, friends, and neighbours often expanded to non-emergency needs such as children's schooling, house building, and farming inputs.

Small *warung*, mostly run by women, were another source of limited, short-term, and interest-free loans. In the villages, there were one or two of these small shops for roughly every dozen houses. Giving a short-term loan was an important service for each *warung*'s regular costumers (*langganan*). Small household items such as rice, cooking oil, sugar, salt, and MSG were goods that were often obtained from a *warung* and paid for a couple of days or weeks later. If they did not provide such credit, the *warung* would lose their *langganan*, while for the *langganan*, not repaying the loan would lead to the loss of an important source of credit.

Another way to acquire cash, especially for women in poor households, was to tap into group savings (*arisan*). Two active women's *arisan* groups in the village — one in the hamlet of Temiangan and another one in Rigis Atas — consisted of 20 or more neighbouring women. In Temiangan, each year after the coffee harvest season, each member deposited an agreed-upon sum of cash. This collective saving was continually accumulated, and by 2002, the group had collected more than Rp 10 million. Rather than agreeing to disburse the accumulated savings, the group decided to continue saving collectively. If life got harder and there were no alternative sources of cash, then the savings would be disbursed. In Rigis Atas, the group had only formed in 2002. The members met twice a month to collect the money and randomly select the recipient of the pooled cash. Recipients used the money for consumption (such as buying rice or paying children's school fees) or productive investments (such as buying a goat, making a start in chilli farming, or adding more stock to a *warung*).

9. Conclusion

The evidence presented in this monograph illustrates the flexibility of smallholder farmers' responses to constraints and opportunities. With attractive export coffee prices over the previous two decades, smallholders in Sumber Jaya and Way Tenong had allocated available labour and capital to intensive robusta coffee farming. Following the drop in coffee prices after the 1997–98 monetary crisis (*krismon*), labour and non-labour inputs to coffee farming were gradually reduced. Although the returns to land from coffee farming decreased, the return to labour remained attractive. Compared with coffee, vegetable farming provided higher production per unit of land and per unit of labour, which resulted in available labour and capital shifting from export crops to intensive vegetable farming for the domestic market.

Diversification in smallholder agricultural production in Sumber Jaya and Way Tenong was made possible because of the significant amount of labour and non-labour inputs acquired through non-market and non-capitalist relations. Access to land could be obtained through borrowing, sharecropping, inheritance, or squatting on forestry land. Relatives, friends, and rotational savings groups often provided interest-free credit. When commercial credit was not available through formal sources, usury was an alternative. Family members, sharecropping arrangements, and reciprocal labour exchange also provided alternatives to paid labour.

Another response to farming constraints was that more households sought off-farm, non-farm, and off-village sources of income. During my fieldwork in Sumber Jaya and Way Tenong, I was told many stories on this topic. A wealthy villager in Gunung Terang bought a shrimp pond in South Lampung; another wealthy villager in Bedeng Sari bought a palm oil garden in Jambi; and a family in Rigis Atas bought a rice field in Bandar Jaya in Central Lampung. When I was completing my fieldwork in early 2003, five men from Bedeng Sari and Buluh Kapur left the village to work overseas in Malaysia, leaving their wives and children behind. Among the poor, I often heard discussions of villagers' plans to send family members to work elsewhere, for example in factories in Java or farms in other parts of Lampung or in the neighbouring provinces of Bengkulu, Jambi, and Riau.

The diversity of options offers a challenge for future research on household economies. '[T]he household', as Rigg (2003: 199) vividly points out, 'has become more fractious, fractured and fragmented'. Members of a household no longer necessarily live continuously under the same roof, which means that consumption and income generating activities are often conducted separately.

Swift responses to constraints and opportunities have been the key strategies for families to guarantee their own welfare. The ethos is that one is expected to stand on one's own. It is everyone's stated goal to have a better income, better education for their children, better housing, and posses more modern household goods. Attaining personal prosperity was a household or family responsibility, and families saw the attainment of this goal as one facilitated through engagement with state-led development initiatives. It was in this context that villagers organised their social lives.

This monograph has examined the ways in which people experience 'development' and the ways they shape and maintain their modes of life. It focuses on the forces that drive changes, their consequences, and the ways people cope with them. Development brings mixed results and effects. Marginality emerges from ongoing relations between centre and periphery, rather than from the resistance of the periphery toward the centre or the absence of centre–periphery relations.

However, state/centre and people/periphery relations do not necessarily lead to marginality and development failures. People in geographically marginal areas position themselves within the orbit of state power in order to promote resource flows from the centre to the periphery, while restricting resource extraction from the periphery to the centre. In dealing with development initiatives and their concomitant changes, people's responses or strategies involve accommodation, collaboration, and compliance, as well as competition and resistance.

Villagers in Sumber Jaya and Way Tenong had transformed the forest frontier into a flourishing region (Chapter Three) and had adapted well to centralised government. In the past, villagers had invited more migrants and created more administrative villages (Chapter Three), turned the region into a pocket of Golkar voters (Chapter Four), and assimilated village leadership into the state framework (Chapters Four, Seven, and Eight) in an attempt to attract state resources to the village. On the other hand, villagers resisted government attempts to transform their smallholder fields into state plantation forests (Chapter Five).

Indonesia's post-New Order *reformasi* and the consequent decentralisation of administration brought changes in the organisation of village life. Some villagers began to collaborate with forestry authorities in 'forest management' through the granting of community forestry contracts (Chapter Six). Villagers no longer aligned themselves with Golkar, and village leaders paid more attention to policies at the district (*kabupaten*) level (Chapters Four and Six). In West Lampung, the term for the administrative village was changed from *desa* to *pekon*; the term for sub-village or hamlet from *dusun* to *pemangku*; the term for village head from *lurah* or *kepala desa* to *pertain*; and for hamlet head from *kepala dusun* to *pemangku*. But despite their name changes, the functions of these units remained much the same.

Reformasi and decentralisation would continue to bring changes in the organisation of village life. More decision-making would perhaps be taken at the provincial and district levels. With decentralisation, villagers in Way Tenong and Sumber Jaya now needed to adopt new strategies. The question for future research on this matter is whether they would be as successful as before.

Reformasi and decentralisation also fostered the presence of NGOs and people's organisations in the villages. State institutions were no longer the only extra-village agencies that villagers engaged with. In Sumber Jaya and Way Tenong, apart from WATALA and ICRAF, there were several other organisations working with villagers. For example, in 2002, the Komite Anti Korupsi (Anti-Corruption Committee), which was based in Bandar Lampung, recruited and trained village leaders for its campaign against corruption and in rights advocacy. This body also conducted an investigation into illegal logging practices in the region and published its findings. Some Semendonese villagers in Sukaraja in Way Tenong formed an organisation called Yayasan Cinta Lingkungan (Caring for the Environment Foundation), whose activities included gaining support for the protection of the Kalpataru forest, a campaign against illegal logging, promoting the formation of community forestry groups in some villages, and obtaining assistance and credit for village cooperatives. Villagers in Dwikora (Bukit Kemuning) joined regional and national farmers' organisations that fought for farmers' land rights. The emerging and diverse issues and opportunities regarding village relations with non-government institutions formed part of the *reformasi* landscape of highland West Lampung.

In a speech to inaugurate various development projects in front of 20,000 people in Central Lampung on 31 August 2004, President Megawati claimed that she was surprised to learn that Lampung was among the three poorest provinces in Indonesia (*Lampungonline*, 31 August 2004). She could understand that East Nusa Tenggara, due to its limited resource potential, might be one of Indonesia's poorest provinces. Lampung, however, was well known as a producer of abundant agricultural commodities such as rice, coffee, pepper, and sugar. It was not supposed to be a poor province. She concluded her speech by asking industries (*pengusaha*) to do more to assist local farmers (*petani*).

President Megawati was correct in pointing out that the absence of a mutually supportive relationship between industries and smallholder farmers was an important issue for a better future for the Lampung people, but these relationships were not the whole story. The historical transformation of Lampung province had been driven largely by colonial and post-colonial central planning and development initiatives (Chapter Two), and it is argued that the results of these initiatives were linked to centre–periphery relations and the emergence of uneven resource flows.

In an interview in 1997, Harris Hasyim, the head of the Lampung Development Planning Office (Badan Perencanaan Pembangunan Daerah), acknowledged that poverty eradication in the province had been slow (*lamban*) (*Angkatan Bersenjata*, 26 November 1997). He attributed this to continued high population growth, and later noted that:

> [d]espite the end of the transmigration program [in Lampung] in 1977–78, the migration of population from Java to Lampung has continued to flow and is very difficult to control. They [the migrants] are poor people from Java … (*Kompas*, 25 June 2001).

The high incidence of poverty in Lampung was perceived to be the result of the growing number of poor migrants from Java. Thus, in contrast to past decades, when the migration of people from Java to Lampung was seen as the source and justification for Lampung's regional development (Chapter Two), the migration of people from Java was now seen as inimical to development and poverty eradication. Viewed as a blessing in the past, the inflow of migrants was now seen as a curse.

Although it is true that latecoming migrants formed the bulk of the poor stratum in Sumber Jaya and Way Tenong, children of earlier migrants and older generations were also found in the village's lowest socio-economic stratum (Chapter Seven). Moreover, not all latecoming poor migrants came directly from Java. Many of them had previously lived in other regions within the province of Lampung. Nonetheless, linking high incidences of poverty with the flow of poor migrants from Java could divert attention from the link between poverty and uneven resource flows in the context of centre–periphery relations.

Geographically, 'poor zones' in Lampung were areas where the inflow of state resources was limited and where natural resources tended to be extracted by central elites (Chapter Two). The poor zones on the plains of Way Kanan, Tulang Bawang, North Lampung, and South Lampung, represented the last of the transmigration areas. Instead of irrigated rice fields, the transmigrants were allotted dry fields to cultivate. Although the land areas allotted to them were larger than the irrigated rice fields granted to transmigrants during the colonial and early post-colonial periods, the fields were not large enough for rotational cultivation. Cultivated without effective fallow periods, poor soils deteriorated further, and incomes were reduced accordingly.

Nonetheless, the opening of these transmigration sites attracted a flow of spontaneous migrants and more administrative villages were created (Chapter Two). Once again, these migrants could only obtain small fields. In the colonial and the early post-colonial eras, transmigration programs were preceded or immediately followed by the construction of irrigation canals and a range of regional and rural development projects (construction of roads and villages, education and health facilities, agricultural extension, and so forth). This was not the case with later post-colonial transmigration programs. In the 1980s, Lampung ceased to be a major transmigration receiving area, and funds from the central government for regional development ceased to flow. Thus, unlike the earlier transmigrants, the last transmigrants and those migrating 'spontaneously' did not secure the expected flow of resources from centre to periphery. Instead, the development of the estate sector stimulated the extraction of resources from these peripheral areas.

Because there were few other options, many smallholder farmers in these poor regions practised intensive farming. Demand from feed industries for maize, and from food industries for cassava and soybean, led to the application of chemical fertilisers and hybrid seeds in maize cultivation. But while production increased, farmers' incomes remained low because of high input costs and price fluctuations.

In contrast, the late colonial and early post-colonial transmigration areas in Lampung were transformed into so-called 'wealthy zones'. Metro, East Lampung, and Central Lampung belonged to this category, with a relatively high incidence of 'wealthy' families. Cultivation of irrigated rice fields drove the economy in these areas, and thanks to the flood of state resources for regional development, the economy of these areas was buoyant. A few decades before, the early colonial transmigration settlements of Pringsewu, Gading Rejo and Gedong Tataan in the district of Tanggamus also belonged to this category. But after two or three generations of land division through inheritance, the number of landless and near-landless villagers grew, and poverty became an issue. Input costs for rice cultivation were high, and despite the floor price set by the government, incomes were low.

In 'wealthy zones' with the cultivation of perennial cash crops such as pepper and coffee, a higher household income enabled smallholders to live above a subsistence level. The population of these 'wealthy zones' consisted of spontaneous migrants as well as natives. The bulk of these smallholders had a better life. However, here too shrinking landholdings and increasing landlessness was an emerging problem (Chapter Seven), although it was emerging at a slower pace compared to irrigated rice field areas. Coping with the fluctuation of prices for cash crops was another perennial issue.

Smallholders in upland Lampung were quick to respond to constraints and opportunities. In the years following the drop in the prices of cassava, maize, and soybean, farmers ceased to plant these crops and planted others instead. Among these other crops, bananas, lemons, and watermelons emerged as commodities that Lampung exported to Java. Surplus production was also achieved for eggs, chickens, goats, and cattle. When tapioca factories and feed industries stimulated a rise in the prices of cassava and maize, hundreds of thousands of hectares of land were replanted with these two crops. A similar pattern emerged amongst smallholder coffee farmers following the drop in coffee prices. In Tanggamus, some farmers planted vanilla and cocoa; in Liwa, smallholders started to plant lemon or other citrus trees in addition to vegetables; while in Sekincau, Way Tenong, and Sumber Jaya, intensive vegetable farming became the alternative. Another response was to seek off-farm, non-farm, and off-village sources of income.

The farmers' decision to diversify their farming and non-farming income sources was linked to the absence of relations between farmers and industries. For decades Lampung had been the home of agribusiness with crops such as oil palm, sugar, tapioca, animal feed, and more recently canned food (pineapple, rambutan, coconut) and shrimp. Most of these industries obtained raw materials from their own plantations. They bought additional raw materials from smallholders, but as farmers frequently complained, the prices they set were too low to provide the farmers with an adequate income. Lampung was also the home of large exporters of coffee, pepper, and copra (dried coconut meat). Like the processing industries, these exporters did not offer much help to the smallholders in their attempts to obtain better incomes.

Conflicts between plantation companies and villagers had become a regular feature of Lampung politics. As Lucas and Warren (2003) have noted, the Indonesian people's struggle for land rights and demand for agrarian law reform remained as unfinished business in post-Suharto Indonesia. For example, in October 2000, about 1,000 people from six villages burned the office, managers' houses, hall, and clinic of the state-owned PT Perkebunan Nusantara in Kalianda, South Lampung (*Kompas*, 4 October 2000). The only building that the villagers did not burn was the company's mosque. Fortunately, all of the company's managers were evacuated prior to the arson attack and no one was injured. The attack was triggered by villagers' anger after a story was published about company guards torturing five villagers accused of stealing coconuts from the plantation. It was reported that, prior to this attack, in mid-2000, farmers demolished 500 hectares of the company's palm oil plantations in Bergen, also in South Lampung, and converted the land into a settlement (*Kompas*, 25 June 2001). In Bunga Mayang, North Lampung, nearly 5,000 hectares of the company's sugarcane plantations were claimed as communal *adat* land by the surrounding native Lampung people. In Tulang Bawang, the *adat* communities claimed 12,800 hectares of land that the government granted to PT SIL, owned by the Salim Group, for sugarcane plantations. In 2001, PT Tris Delta, a Taiwanese company developing a pineapple plantation and canning factory in Central Lampung, was closed after the company's land was taken over by thousands of transmigrants from adjacent villages.

Assistance given to coffee smallholders was an example of the rift that existed between what exporters (and the government) offered and what farmers expected. President Megawati's visit to Lampung on 31 August 2004 was to declare Lampung as a 'national coffee *étalase*' (showcase). President Megawati also noted that she expected Lampung, the producer of 60–65 per cent of national coffee production, to lead the nation in coffee research and development. In 1999, the Asosiasi Exportir Kopi Indonesia (Indonesian Coffee Exporters Association) opened its Centre for Coffee Extension and Development (Pusat Penyuluhan dan Pengembangan Kopi) in Hanakau, West Lampung. The centre had 10 hectares of exemplary gardens (*kebun percontohan*) for smallholders where it demonstrated the cloning, cultivation techniques, and post-harvest handling procedures needed to produce a higher yield and quality of coffee (*Lampung Post*, 24 August 2004). The smallholders would have adopted this advice if the price had been attractive, but with a decline in coffee prices in recent years, they chose to reduce labour and non-labour inputs in coffee farming and diversify their crops and income sources.

To suggest an absence of collaborative relationships between industries and farmers in Lampung is misleading. The problem more directly involved the nature and scale of these relationships. For example, PT Nestle Beverage, a multinational food company whose factory in Panjang, Bandar Lampung, produces Nescafé instant coffee for the world market, provided assistance for the provision of inputs, cultivation techniques, post-harvest handling, and marketing to a village cooperative in Pulau Panggung, Tanggamus (*Lampung Post*, 16 August 2004). In a similar fashion, the government promoted partnerships between feed industries, banks and smallholder maize farmers. Contract farming was also gaining recognition in Lampung (Chapter Two), but experience with this arrangement had so far been disappointing. In the late-1970s, contract farming in Lampung began with rubber on the North Lampung plain. It had been more recently adopted in sugarcane, oil palm, and shrimp production. Apart from land conflicts, a common problem that rose from the implementation of these schemes was that farmers felt that they were being exploited. In May 1999, 4,000 shrimp pond smallholders from Dipasena, Tulang Bawang, demonstrated and camped overnight at the governor's office in Bandar Lampung. PT Dipasena was owned by a tycoon named Syamsul Nursalim of the Gajah Tunggal group. By mid-October, the number of smallholders who had joined the demonstration and the encampment at the governor's office had doubled. Their demands were for a revision of the contract, making it more beneficial to farmers, and the curtailment of intimidation and unilateral contract termination by the company (*Kompas*, 15 October 1999). The conflict continued and the industry collapsed. The company's assets were later handed to the Indonesian Bank Restructuring Agency (IBRA), a new negotiation with smallholders commenced, and in early 2004, shrimp production was resumed (*Tempointeraktif*, 4 May 2004). In September 2004, 200 of 1,600 smallholders engaged in contract farming with PT BNIL on a palm oil plantation accused the company of not paying their entitlements from the harvests (*Lampung Post*, 2 September 2004). According to media reports, exploitation emerged as a point of contention in contract farming, but more information on other cases would need to be reviewed for a general conclusion to be made.

About 1 million hectares of Lampung's territory was classified as state forest (*kawasan hutan negara*). Smallholder farmers controlled a significant portion of these lands. To date the forestry authorities and the industry had shown a great desire to own the trees and land and force the villagers to become gatherers of minor forest products at best, or a cheap labour force at worst, leading villagers to relentlessly resist these attempts (Chapter Five). This was particularly true of the production forest zones (*hutan produksi*), all of which were legally controlled either by the state-owned forestry company (PT Perhutani) or by private companies. In protection forest zones (*hutan lindung*) under the jurisdiction of provincial and district governments, the forestry authorities hesitantly began to invite villagers to collaborate in 'forest management'. In Sumber Jaya and Way Tenong (Chapter Five), permission to continue coffee farming was given to community groups that were expected to plant more trees in their gardens and protect the remaining forests. Another example comes from Tanggamus, where reforestation funds had been given to 15 community groups that were responsible for planting trees in their coffee gardens (*Lampung Post*, 13 August 2004), instead of this being done by a private company, the military, or another government agency. The eviction of forest squatters had continued, but the ways in which it was carried out were different. In August 2004, over 2,000 farmers living inside Gunung Betung Provincial Forest Park (Taman Hutan Raya) in South Lampung were asked to dismantle their huts and move elsewhere (*Lampungonline*, 24 August 2004). Unlike in the past, this time there was no violence, no arson, and no crop demolition. More importantly, the farmers were officially forbidden but discreetly allowed to continue farming coffee and cocoa inside the park.

The two national parks in the province (Way Kambas in East Lampung and Bukit Barisan Selatan in West Lampung and Tanggamus) were faced with serious difficulties which could result in wildlife disappearing and being replaced by smallholder fields. The problems included illegal hunting and poaching, elephant attacks on surrounding villages, and the expansion of smallholders' fields. Coordination between the park authority and local government was minimal, and coordination with villagers was even more limited. Unless coordination between these three groups were to improve, the wildlife in these two parks would be under serious threat.

The changes in the management of protection forests (and the absence of change in production and conservation forests) were brought about by post-Suharto decentralisation in natural resource management. While the management of protection forests was carried out by local governments, the management of production forests was in the hands of companies, and conservation forests in those of the central government (the Ministry of Forestry). Interactions between the state, local people, and natural resources would thus remain as a vital issue on the provincial development agenda.

References

Antlov, H., 1995. *Exemplary Centre, Administrative Periphery: Rural Leadership and the New Order in Java*. Surrey: Curzon Press.

Bastin, J.S., 1965. *The British in West Sumatra (1685–1825)*. Kuala Lumpur: University of Malaya Press.

Benoit, D., 1989. 'Migration and Structures of Population.' In M. Pain (ed.), *Transmigration and Spontaneous Migrations in Indonesia: Propinsi Lampung*. Jakarta: Departemen Transmigrasi [Department of Transmigration]. Paris: ORSTOM.

Boserup, E., 1965. *The Conditions of Agricultural Growth: The Economics of Agrarian Change under Population Pressure*. Chicago: Aldine.

Breman, J., 1982. 'The Village on Java and the Early Colonial State'. Journal of Peasant Studies 9: 189-240.

Brookfield, H., 2000. *Exploring Agrodiversity*. New York: Columbia University Press.

————,2001. 'Intensification, and Alternative Approaches to Agricultural Change.' *Asia Pacific Viewpoint* 42: 181–92.

Brown, M.F., 1996. 'On Resisting Resistance.' *American Anthropologist* 98: 729-35.

Bulbeck, D., A. Reid, L.C. Tan and Y. Wu, 1999. *Southeast Asian Exports Since the 14th Century: Clove, Pepper, Coffee and Sugar*. Singapore: Institute of Asian Studies.

DPU (Departemen Pekerjaan Umum), 1995. 'Bendungan Besar di Indonesia [Big Dams in Indonesia].' Jakarta: DPU.

Dove, M.R.,1986. 'The Ideology of Agricultural Development in Indonesia.' In C. MacAndrews (ed.), *Central Government and Local Development in Indonesia*. Singapore: Oxford University Press.

Dove, M.R. and D.M. Kammen, 2001. 'Vernacular Models of Development: An Analysis of Indonesia under the "New Order".' *World Development* 29: 619–639.

Elmhirst, R.J., 1997. Gender, Environment, and Culture: A Political Ecology of Transmigration in Indonesia. London: University of London (Ph.D. thesis).

Escobar, A., 1995. *Encountering Development: The Making and Unmaking of the Third World*. Princeton (NJ): Princeton University Press.

Esman, M.J. and N.T. Uphoff, 1984. *Local Organization: Intermediaries in Rural Development*. Ithaca (NY): Cornell University Press.

Ferguson, J., 1994. *The Anti-Politics Machine: 'Development', Depoliticization, and Bureaucratic Power in Lesotho*. Minneapolis (MN): University of Minnesota Press.

Green, M., 2000. 'Participatory development and the appropriation of agency in Southern Tanzania.' *Critique of Anthropology* 20: 67–89.

Grillo, R.D., 1997. 'Discourses of Development: The View from Anthropology.' In R.D. Grillo and R.L. Stirrat (eds), *Discourses of Development: Anthropological Perspectives*. Oxford and New York: Berg.

Hadikusuma, H., 1989. *Masyarakat dan Adat Budaya Lampung [Community and Customs of Lampung]*. Bandung: Mandar Maju.

Hardjono, J.,1977. *Transmigration in Indonesia*. Kuala Lumpur: Oxford University Press.

Hart, G., A. Turton and B. White (eds), 1989. *Agrarian Transformations: Local Processes and the State in Southeast Asia*. Berkeley (CA): University of California Press.

Heeren, H.J., 1979. *Transmigrasi di Indonesia [Transmigration in Indonesia]*. Jakarta: PT Gramedia.

Hefner, R., 1990. *The Political Economy of Mountain Java: An Interpretative History*. Berkeley (CA): University of California Press.

Hobart, M., 1993. 'Introduction: The Growth of Ignorance.' In M. Hobart (ed.), *An Anthropological Critique of Development: The Growth of Ignorance*. London: Routledge.

Jaspan, M.A., 1964. From Patriliny to Matriliny: Structural Change among the Redjang of Southwest Sumatra. Canberra: Australian National University (Ph.D. thesis).

Kahn, J.,1999. 'Culturalising the Indonesian uplands.' In T.M. Li (ed.), *Transforming the Indonesian Uplands: Marginality, Power and Production*. Singapore: Harwood Academic Publishers.

Kingston, J.B., 1987. The Manipulation of Tradition in Java's Shadow: Transmigration, Decentralization, and the Ethical Policy in Colonial Lampung. New York: Columbia University (Ph.D. thesis).

LeBar, F.M., 1976. *Insular Southeast Asia: Ethnographic Studies – Section 1: Sumatra*. New Haven (CT): Human Relations Area Files.

Levang, P., 1989. 'Farming systems and household incomes'. In M. Pain (ed.), *Transmigration and Spontaneous Migrations in Indonesia: Propinsi Lampung*. Jakarta: Departemen Transmigrasi [Department of Transmigration]. Paris: ORSTOM.

Li, T.M., 1999a. 'Introduction.' In T.M. Li (ed.), *Transforming the Indonesian Uplands: Marginality, Power and Production*. Singapore: Harwood Academic Publishers.

————, 1999b. 'Marginality, Power, and Production: Analysing Upland Transformation.' In T.M. Li (ed), *Transforming the Indonesian Uplands: Marginality, Power and Production*. Singapore: Harwood Academic Publishers.

————, 2001. 'Relational Histories and the Production of Difference on Sulawesi's Upland Frontier.' *Journal of Asian Studies* 60: 41–66.

————, 2002. 'Local Histories, Global Markets: Cocoa and Class in Upland Sulawesi'. *Development and Change* 33: 415–437.

Lucas, A. and C. Warren, 2003. 'The State, the People, and Their Mediators: The Struggle over Agrarian Law Reform in Post-New Order Indonesia.' *Indonesia* 76: 87-125.

McKinnon, E.E., 1993. 'A Note on Finds of Early Chinese Ceramics Associated with Megalithic Remains in Northwest Lampung.' *Journal of Southeast Asian Studies* 24: 227-238.

Michon, G., H. de Foresta, P. Levang, and A. Kusworo, 2000. 'Repong di Pesisir Krui, Lampung [Tree-Gardens in Pesisir Krui, Lampung].' In H. de Foresta, A. Kusworo, G. Michon and W.A. Djatmiko (eds), *Ketika Kebun Berupa Hutan: Agroforest Khas Indonesia [When Gardens Become Forests: Indonesia's Unique Agroforests]*. Bogor: International Centre for Research in Agroforestry.

Netting, R.M., 1993. *Smallholders, Householders: Farm Families and the Ecology of Intensive, Sustainable Agriculture*. Stanford (CA): Stanford University Press.

Ortner, S.B., 1984. 'Theory in Anthropology since the Sixties.' *Comparative Studies in Society and History* 26: 126–166.

————, 1995. 'Resistance and the Problem of Ethnographic Refusal.' *Comparative Studies in Society and History* 37: 173–195.

Pain, M., 1989. 'Spatial Organisation and Regional Development.' In M. Pain (ed.), *Transmigration and Spontaneous Migrations in Indonesia: Propinsi Lampung.* Jakarta: Departemen Transmigrasi [Department of Transmigration]. Paris: ORSTOM.

Peluso, N.L., P. Vandergeest and L. Potter, 1995. 'Social Aspects of Forestry in Southeast Asia: A Review of Postwar Trends in the Scholarly Literature.' *Journal of Southeast Asian Studies* 26: 196-218.

Pelzer, K.J., 1945. *Pioneer Settlement in the Asiatic Tropics: Studies in Land Utilization and Agricultural Colonization.* New York: American Geographical Society.

Perdana, A.A. and J. Maxwell, 2004. 'Poverty Targeting in Indonesia: Programs, Problems and Lessons Learned.' Jakarta: Centre for Strategic and International Studies.

Pigg, S.L., 1992. 'Inventing Social Categories through Place: Social Representations and Development in Nepal.' *Comparative Studies in Society and History* 34: 491–513.

Potter, L., 2001. 'Agricultural Intensification in Indonesia: Outside Pressure and Indigenous Strategies.' *Asia Pacific Viewpoint* 42: 305-324.

Potter, L. and S. Badcock, 2004. 'Tree Crop Smallholders, Capitalism, and Adat: Studies in Riau Province, Indonesia.' *Asia Pacific Viewpoint* 45: 341–356.

Quarles van Ufford, P. (ed.), 1987. *Local Leadership and Programme Implementation in Indonesia.* Amsterdam: Free University Press.

Rigg, J., 2003. *Southeast Asia: The Human Landscape of Modernization and Development.* London: Routledge.

Ruiter, T.G., 1999. 'Agrarian Transformations in the Uplands of Langkat: Survival of Independent Karo Batak Rubber Smallholders.' In T.M. Li (ed.), *Transforming the Indonesian Uplands: Marginality, Power and Production.* Singapore: Harwood Academic Publishers.

Sachs, W. (ed.), 1992. *The Development Dictionary: A Guide to Knowledge as Power.* London: Zed Books.

Sakai, M., 1999. The Nut Cannot Forget Its Shell: Origin Rituals among the Gumai of South Sumatra. Canberra: The Australian National University (Ph.D. thesis).

Schrauwers A., 1999. '"It's Not Economical": The Market Roots of a Moral Economy in Highland Sulawesi.' In T.M. Li (ed.), *Transforming the Indonesian Uplands: Marginality, Power and Production*. Singapore: Harwood Academic Publishers.

Scott, J.C., 1985. *Weapons of the Weak: Everyday Forms of Peasant Resistance*. New Haven (CT): Yale University Press.

———, 1998. *Seeing Like a State: How Certain Schemes to Improve the Human Condition Have Failed*. New Haven (CT): Yale University Press.

Sevin, O., 1989. 'History and Population.' In M. Pain (ed.), *Transmigration and Spontaneous Migrations in Indonesia: Propinsi Lampung*. Jakarta: Departemen Transmigrasi [Department of Transmigration]. Paris: ORSTOM.

Sukendar, H., 1979. *Laporan Penelitian Kepurbakalaandaerah Lampung [Report on Archaeological Research in Lampung]*. Jakarta: Proyek Penelitian dan Penggalian Purbakala Departemen Pendidikan dan Kebudayaan [Archeological Excavation Research Project of the Ministry of Education and Culture].

Suryanata, K., 1999. 'From Home Gardens to Fruit Gardens: Resource Stabilisation and Rural Differentiation in Upland Java'. In T.M. Li (ed.), *Transforming the Indonesian Uplands: Marginality, Power and Production*. Singapore: Harwood Academic Publishers.

Tjondronegoro, S.M.P., 1984. *Social Organization and Planned Development in Rural Java*. Singapore: Oxford University Press.

Tsing, A.L., 1993. *In the Realm of the Diamond Queen: Marginality in an Out-of-the-Way Place*. Princeton (NJ): Princeton University Press.

Utomo, K., 1975. *Masyarakat Transmigran Spontan di Daerah Wai Sekampung, Lampung [Spontaneous Transmigrant Communities in Wai Sekampung, Lampung]*. Yogyakarta: Gadjah Mada University Press.

Vandergeest, P. and N.L. Peluso, 1995. 'Territorialization and State Power in Thailand.' *Theory and Society* 24: 386–427.

Warren, C., 1993. Adat *and Dinas: Balinese Communities in the Indonesian State*. Kuala Lumpur: Oxford University Press.

Wertheim, W.F., 1959. 'Sociological Aspects of Inter-Island Migration in Indonesia'. *Population Studies* 12: 184–201.

White, B., 1989. 'Problems in the Empirical Analysis of Agrarian Differentiation.' In G. Hart, A. Turton and B. White (eds), *Agrarian Transformations: Local Processes and the State in Southeast Asia*. Berkeley (CA): University of California Press.

White, B., and G. Wiradi, 1989. 'Bases of Inequality in Javanese Villages.' In G. Hart, A. Turton and B. White (eds), *Agrarian Transformations: Local Processes and the State in Southeast Asia*. Berkeley (CA): University of California Press.

Wolf, E.R., 1982. *Europe and the People without History*. Berkeley (CA): University California Press.

www.ingramcontent.com/pod-product-compliance
Lightning Source LLC
Chambersburg PA
CBHW061217270326
41926CB00028B/4677